Rick Fish

RELIABILITY
for the
TECHNOLOGIES

Second Edition
RELIABILITY
for the
TECHNOLOGIES

Leonard A. Doty

An ASQC Quality Press Book
published by
Industrial Press Inc.

Library of Congress Cataloging-in-Publication Data

Doty, Leonard A.
  Reliability for the technologies / Leonard A. Doty. — 2nd ed.
    p.   cm.
  "An ASQC quality press book."
  Bibliography: p.
  Includes index.
  ISBN 0-8311-1169-0
  1. Reliability (Engineering)   I. Title.
TA169.D68   1989
620'.004252--dc20   89-15282
                                                        CIP

RELIABILITY FOR THE TECHNOLOGIES, SECOND EDITION

INDUSTRIAL PRESS INC.
200 Madison Avenue, New York, New York 10016-4078

Composition by Edwards Brothers, Inc.,
Ann Arbor, Michigan

4 6 7 5 3

To My Mother
A model of quiet strength, unselfish love, and unrelenting devotion
to her ideals.

CONTENTS

PREFACE TO THE SECOND EDITION

The purpose of this second edition to *Reliability for the Technologies* is to clarify further some of the concepts presented in the first edition and to add concepts developed since the original publication of this text. Changes or additions or both have been made to Chapters 1, 4, 6, 8, 9, and 12. Some of the chapter titles have been changed to reflect more closely the nature of the material presented therein.

In Chapter 1, the bathtub analysis has been expanded to include two limits; one limit relating to design requirements and the other limit to repair and maintenance. In addition, the concepts and procedures relating to failure rate and mean life calculations were moved to this chapter from Chapter 6. Chapter 4 now includes a section showing the derivation of the single-term binomial formula from basic probability theory. Chapter 6 has had many changes in addition to moving the failure rate calculations to Chapter 1. The use of the Poisson distribution to determine the probability of more than one failure was found to be in error and has been omitted. Four distributions are now used in Chapter 6 to determine component reliability, and a better method of comparing their respective results has been instituted. These same four distributions have been used in Chapter 8 to determine confidence limits, and the same comparison procedures were continued. The calculation of confidence limits has also been expanded to a two-step procedure, using the basic reliability formulas from Chapter 6 along with the confidence-limit formulas, so that a lower confidence limit can be calculated from basic requirements rather than just assumed. Similar procedures, to those used in Chapters 6 and 8, have also been used in Chapter 9 to calculate maintainability and maintainability confidence limits, except that the log-normal has been added as the basic maintainability distribution. Finally, a section on failure rate allocation was added to Chapter 12.

In recent years, the Weibull distribution has become increasingly accepted by reliability engineers as the basic reliability distribution. One of the problems with using the Weibull, however, is its dependence on an enormous amount of tables and on the use of graphical-solution techniques. Using graphs to solve reliability problems has two fundamental weaknesses. First, the graphical technique is limited to solving for only two of the three reliability questions described in Chapter 6 and only two of the four confidence limit questions described in Chapter 8. Second, graphical solutions are inherently imprecise; it is just too easy to make errors in reading the graph and in fitting a line to the points. Although this text presents the Weibull graphical-solution techniques, formulas have also been developed, and presented, for solving these problems using mathematical analysis without the use of graphs. However, the Weibull confidence limit formulas used in this text have been empirically derived, rather than by rigorous mathematical analysis, and should be used with care.

I wish to thank my colleagues at Weber State College and ASQC (American Society for Quality Control), especially Mr. Roy Thornock, Mr. James Puggsley, and Mr. Kenneth Jensen, for their valuable assistance in developing these procedures and formulas. I also wish to thank Mr. Gordon Griffin for his valuable assistance in checking the manuscript for errors

PREFACE
TO
THE
FIRST
EDITION

The complexity and accelerated growth of technology, as well as modern safety and liability laws, have increased the need for more reliable products. Therefore, reliability in technology has become more important in this complex, modern world. More than ever before, an understanding of how reliability functions is needed by all who are involved in any aspect of product life.

The purpose of this volume is to promote an understanding of reliability, even (and especially) among those who have limited, specialized knowledge. Therefore, *Reliability for the Technologies* presents the subject in a simple style, free of the complexities of calculus; all that is required of the reader is a thorough grounding in algebra. Since a knowledge of algebra is so necessary to the understanding of the subject matter of this volume, a review of the more important algebraic rules and procedures is presented in the Appendix.

Reliability for the Technologies begins with an overview of reliability and then proceeds to discuss the important probability rules that will be needed. The normal, binomial, and Poisson probability distributions are presented prior to any discussion of reliability, since these distributions are critical ingredients of reliability theory. The more complex Weibull and log-normal distributions are discussed briefly as part of later chapters, but they are not needed to understand the basic reliability theory as presented here. One other probability distribution, the chi-square, is discussed in Chapter 8 in relation to confidence limits. Practical examples are used at every stage to illustrate the concepts introduced.

After the probability foundation is laid (Chapters 2 through 5), the book continues with the heart of reliability theory. Chapters 6, 7, and 8 explain failure analysis and calculation of individual component and system reliabilities, reliability prediction of complex systems, and determination of confidence limits for the control of production.

In Chapter 9, some of the fundamentals of maintainability and availability are discussed. It is at this time that the log-normal probability distribution is introduced. (Maintenance repair rates have been determined to be mostly log-normal in shape.) Chapters 10 and 11 then explain the procedures for developing sampling plans and control charts for acceptance sampling.

Chapter 12 is the last chapter that relates mainly to reliability. This chapter explains the relationships among reliability, design, and the highly acclaimed FMECA analysis model (failure modes, effects, and criticality analysis). An example is analyzed in detail to illustrate the techniques.

Chapters 13, 14, and 15 deal with the related subjects of safety reliability management, and product liability.

I am indebted to many colleagues for their encouragement and assistance, especially to Dr. Dale Besterfield whose excellent examples contributed greatly to the writing style and content of this volume, and to Dr. John McLuckie and Fred Meyers who contributed much to the algebra review section. Thanks are also due to Mrs. Linda Thornock who typed most of the manuscript.

I am especially indebted to my wife, Lucy, who typed many revisions and who, in the process, taught me to use our word processor.

CHAPTER 1 | INTRODUCTION

The purpose of this chapter is to give an introductory overview of the subject of reliability.

Objectives

1. Define reliability and explain the four elements that make up reliability.
2. Explain the differences between quality defects and reliability defects.
3. Have a general knowledge of some of the design methods used to improve reliability.
4. Have a general knowledge of the important reliability measures.
5. Know some of the manufacturing practices that affect reliability.
6. Understand the difference between "ideal" and "satisfactory" reliability, including the purpose of reliability and the need for a reliability policy.

1.1 Definition of Reliability

Technology has brought untold riches to the modern world, but at the cost of increasing complexity and frustration. Throughout most of history, products have been simple, easily understood, and easily controlled by the individual. Many, if not most, of the modern products, on the other hand, are complex constructions of many interrelating parts, take years of study to understand, and must be controlled through complex organizational structures of many individuals working in unison. This complexity, along with the high cost of modern materials and labor, have created a new problem for society: how to prolong, predict, and guarantee product life. This is the object of reliability, which can be roughly described as quality over the long run.

Reliability has been defined as "the probability of a device performing its purpose adequately for the period of time intended under the stated operating conditions." There are four elements to the definition. First, probability refers to the chance, or likelihood, that the device will work properly. In fact, the terms chance or likelihood can be used as synonyms to probability. Probability is measured as a decimal ratio, between 0 and 1, and is usually expressed as a per cent (the ratio multiplied by 100). Since probability is the chance that something will occur (chance of a success or a failure, depending on what is desired), it is calculated as the ratio of the successes to the total number of possible occurrences or trials.

The second element to the definition is adequate performance. In order to determine whether or not the performance is indeed adequate, a standard is needed. The standard must define what is meant by adequate and must contain effective measurement criteria for comparing actual performance to the standard. Since nothing in real life is ever perfect, the standard, in order to be realistic, must include an allowed variation. This variation is usually expressed, in measurement terms, as plus or minus a variance from the ideal measurement. The total variation, from high to low, is known as the tolerance. Any measurement that falls within this tolerance is considered to be acceptable; it is a success. The actual performance, then, is measured and compared to a standard. If the actual measurement falls within the standard tolerance (2.05 ± 0.02 in., for instance), the performance of the unit is acceptable.

Good standards are written, and written standards are known as specifications. It is important that specifications (often called specs) be understandable by everyone who uses them — from engineers to machine operators, and from college-trained to relatively unlearned personnel. Therefore, a series of rules and a general format for writing specifications have evolved. The first, and by far the most important, rule is that a spec must be specific; "a general spec is no spec at all." A spec must be so clear and understandable that there is no doubt by anyone reading it (especially those who need to read it) as to what is required. A satisfactory reliability specification must:

1. State exactly what is wanted — clearly, definitely, and completely.
2. Provide the means for testing, and explain the methods and procedures, including sampling and computations.
3. Avoid nonessential quality restrictions that add to cost without adding to utility.
4. Conform, as far as possible, to the established commercial standards.

5. Explain where the product is to be used and under what conditions.
6. Contain a statement of the purpose of the product as a guide to usage and a precaution against misapplication.
7. Explain the inspection and testing procedures to be used in determining conformance to standards, including the instruments and personnel involved.
8. State the applicable standards, including tolerances.
9. Contain a statement of the time frame that applies to the product (expected product life).
10. Define failures in terms of product use and explain how they are to be measured.
11. State the maintenance procedures and conditions.

The expected or desired operating time is the third element of the definition. Nothing lasts forever and nothing can perform adequately forever. Therefore, adequate performance must be defined in terms of a time frame; the standard must include a time limit (for example, the unit must operate within 90% of rated capacity for no less than 100 hr). In reliability, there are several different types of time frames: total test time, test period, mean test time, mean repair time, allowed repair time, and mission time. The desired operating time, as far as the definition of reliability is concerned, is the mission time. Mission time is determined by engineering or management and in accordance with system requirements. The total test time is the sum of the individual unit test times and is determined by actual measurement of one or more test samples. The test period is the actual elapsed time of the test. If 10 units are tested for 10 hr each, the test period is 10 hr and the total test time is 100 hr (10 × 10 = 100). The mean test time is the average of several units actually tested and measured. The mean repair time is the average time it actually took to repair the failures. Allowed repair time is the desired time to repair a failure, determined by engineering or management. Note that mission time and allowed repair time are the only times determined by engineering; all the other times are determined by sample measurements.

The final element of the definition is the operating conditions. An article may perform adequately in one set of conditions and quite poorly in another. A part designed for ambient temperature (room temperature), for instance, may be totally inadequate in arctic climate. Environmental conditions (air pressure, temperature, humidity, etc.) must also be included in the reliability standard. An example of a reliability statement, utilizing all four of the definition elements, might be: The part has a 95% probability of operating at greater than 90% of rated capacity for 100 hr without failure at ambient temperature (70 ± 10 degrees) with no more than 60% humidity in a dust-free atmosphere.

1.2 Defects

There are two major types of defects: quality defects and reliability defects. Since quality is a major ingredient of reliability, it is important to understand its effects. Quality is defined as the degree of conformance to specification and/or workmanship standards. Specifications are written by engineering, while workmanship standards are just generally accepted criteria of excellence and are usually unwritten. Quality does not include a time frame, while reliability does.

Quality defects are usually classified into critical, major, and minor. Critical defects will always affect the product's usability. Major defects may, or probably will, affect usability. Minor defects, usually cosmetic defects, will not affect the usability of the product. Cosmetic defects may affect the salability of the product, but not its usability.

Reliability defects, on the other hand, include a time frame. Thus, reliability defects are associated with failures, or the inability of the part to continue to perform its intended function. There are two classifications of reliability defects: catastrophic and wear-out.

Catastrophic defects usually occur randomly; they are the chance defects that cause a sudden failure to occur without warning. These defects are usually quite costly, since they can cause emergency shutdowns of systems at the most inappropriate times. Wear-out failures occur slowly and frequently signal when failure is imminent. This type of slow degradation of quality is often an indication for replacement before a major failure can occur.

1.3 Design Methods for Reliability

The basic objective of reliability design is to meet the product's operational goals. This subject will be covered in more depth in a later chapter, but as an introduction, there are eight fundamental reliability design methods open to the design engineer:

1. Many books and tables of design criteria are available in the various disciplines. They can supply the information needed to design a product.
2. Other similar products can be used for comparison. Seldom does a design stand alone without some application to other designs, even if the comparison is only slight.
3. At times, with new materials and with new applications of old materials, some amount of testing must be done to determine the material or product capabilities.
4. Statistical prediction techniques are extremely helpful, especially with complex products, in determining the probability of meeting the design criteria.

5. Sometimes it is impossible to meet the reliability goals with only one component to do the job. The concept of redundancy — using more than one part to perform the same operation — can then be used to increase the probability of failure-free operation for the specified time.

6. Nothing can last forever, so, in long mission times, the repair and replacement of failed parts can be of critical importance. This "maintenance" function can be assisted by designing for ease of failure identification and replacement. Here the plug-in component can be extremely helpful in increasing the system availability.

7. Special built-in test equipment can be designed into many products to identify and predict failures. These include go—no go gages, and press-to-test circuits for built-in, routine, sequential maintenance checks. These checks can often identify problem areas where potential failures can be replaced before they occur.

8. Debugging is the process of running a part, usually at less than the rated capacity, for a short time, in order to weed out early failures (also called derating). There are three stages of failures in the life of a product: break-in stage, operating stage, and wear-out stage (see Fig. 1.1). During the break-in stage, the failure rate (failures per hour) is relatively high. This is the reason for debugging — to remove the relatively high break-in failures before submitting the parts to normal operation. During the break-in stage, the failure rate decreases until it reaches its lowest point, at which time it remains constant for most of its operating life. Finally, the failures begin to increase again as the parts start to wear out. Technically, these three stages each require three different types of probability distributions (normal, Poisson, and Weibull). In actual practice, however, the Poisson is frequently used for all three.

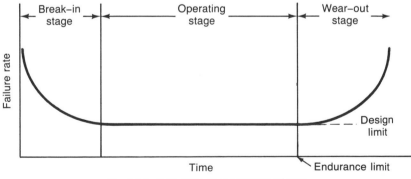

Fig. 1.1 Failure rate as a function of age.

The design limit shown on Fig. 1.1 refers to the lower limit on the failure rate, where the bathtub curve bottoms out. The failure rate — the number of failures per unit of time — can, to some extent, be designed into the product. The early failure rate — the break-in failures or infant mortality — is a function of production, to a great extent, and so cannot be designed into a product. The wear-out stage failure rate, because of the many variables affecting it, is also difficult, if not impossible, to design into the product. Therefore, this design limit actually refers to the operating stage only. It is comparable to the quality control concept of design quality, where design engineering deliberately designs to a particular quality level for a particular product. The endurance limit shown on the graph (Fig. 1.1) refers to the length of time a product will probably last before it begins to fall apart (wear out).

Suppose a product can be designed on several levels of quality, like a car, so that the highest level quality product has very few failures per unit of time (low failure rate) and the lowest quality has many. The design limit would be a measure of this quality. The endurance limit, on the other hand, refers to how long this product will last before repairs are no longer practical, before it can no longer be repaired to its original level of quality (comparable to how long a car will last before it must be scrapped).

Obviously, there is some relationship between the two limits. A product built for low failure rate is likely to outlast another built to a higher failure rate, all other things constant. Unfortunately, this relationship is very variable and difficult, if not impossible, to quantify. Witness the cheaper Japanese cars of the 1970s outlasting most of the more expensive American cars, even many of the most prestigious.

1.4 Measures of Reliability

There are three basic measurements associated with reliability: failure rate, mean life, and probability of survival. In measurements of this nature, numbers are frequently used without identifying the units they represent. This is done for convenience and because the analyst doing the work almost certainly knows if he or she is measuring inches, hours, square feet, etc. It is important to remember, however, that numbers alone are meaningless. They must be associated with units, operating conditions, etc., either specifically (actually written) or inherently (understood).

Failure rate refers to failures per unit of time (1 hr, 100 hr, 1000 hr, etc.). The symbol used for failure rate is λ, the Greek letter lambda. There is a definite relationship between failure rate and mean life. Mean life, variously known as m, $\overline{m}$, MTBF, and MTTF, is the reciprocal of the failure rate λ, and is measured in hours. (m is the general symbol for mean life, $\overline{m}$ is the symbol for the mean of a sample. MTBF is an acronym for mean time between failures, and MTTF stands for mean time to failure of a system where no failures can be tolerated.)

The probability of survival is equal to the probability of zero failures and is the real measure of reliability. Failure rate and mean life are both used to calculate the probability of survival, known as P_s, R, R_s, and sometimes just P. As the mean life increases, the probability of survival increases. However, with failure rate, the situation is reversed. As failure rate increases, P_s decreases, showing the reciprocal relationship between failure rate and mean life ($m = 1/\lambda$).

Failure Rate/Mean Life Calculations

Failure rate is defined as the number of failures per unit of time (minutes, hours, miles, actuations, cycles, etc.), while mean life is time per failure. Thus, failure rate and mean life are just reciprocals of each other (a failure rate of two failures per hour is a mean life of $\frac{1}{2}$ hr per failure; a failure rate of 0.02 failure per hour is a mean life of $1/0.02 = 50$ hr per failure; etc.). Therefore, a formula for one is basically the formula for the other, except for the reciprocal relationship.

The symbol for failure rate is λ, while the symbol for mean life is m. Actually m is just the generic, or universe, symbol for mean life. There are several others, each used in a slightly different way.

1. $\overline{m}$ is used as the mean of a test where actual sample data are available.
2. MTBF means mean time between failures and is used for those units or systems where failures can be allowed and repairs are possible. It can take the place of either the generic m or the sample $\overline{m}$, but is usually used in connection with sample information.
3. MTTF or MTFF mean mean time to failure or mean time to first failure. These terms are used interchangeably for those units or systems where no failure can be tolerated and repairs are not possible (aircraft in flight, missiles, etc.). These terms can also take the place of either the generic m or the sample $\overline{m}$. The context in which they are being used must often be carefully studied to determine which meaning is intended.

The formula for failure rate is

$$\lambda = f/(\Sigma t)$$

where

λ = failure rate

f = number of failures in a test

t = time to failure for a single unit. If the test is time-terminated — the test ends before all units fail — t for the nonfailed units is the test period

Σ = summation sign

The formula for mean life is just the reciprocal of the above formula for failure rate:

$$\overline{m} = (\Sigma t)/f$$

$$m = \overline{m} = \text{MTBF} = \text{MTTF} = \text{MTFF} = 1/\lambda$$

Although the preceding basic formulas are simple, the determination of the total test time (Σt) may sometimes present a problem owing to the different types of tests used. There are four different types of tests.

1. The first type of test is a time-terminated test where failures are immediately repaired and continued on test. The total test time, Σt, in this case, is the test period t times the sample size n (nt).

Example 1.1 Forty units were tested for a test period of 50 hr. At the end of the test there had been five failures that had each been immediately repaired and continued in the test. Find the (1) failure rate and (2) mean life.

Solution:

(1) The failure rate is:

$$\lambda = f/nt = 5 / 40(50) = 0.0025 \text{ failures per hour}$$

(2) The mean life is

$$\overline{m} = nt/f = 40(50)/ 15 = 400 \text{ hours per failure}$$

$$\overline{m} = 1/\lambda = 1/0.0025 = 400 \text{ hours per failure}$$

2. The second type of test is a time-terminated test where the failures are not repaired and replaced but failure times, of the failed units, are logged. The total test time, Σt, in this case, is the sum of the failure times (for the failed units) plus the remaining units (those that did not fail) times the test period. The total test time, then, is $\Sigma t_f + (n - f)t$.

Example 1.2 Assume that the five failures of Example 1.1 are not repaired and replaced and that the failure times are 1, 12, 15, 29, and 43 hr. Find the (1) failure rate and (2) mean life.

Solution:

(1) The failure rate is

$$\lambda = f/[\Sigma t_f + (n - f)t]$$

$$\lambda = 5/[100 + (40 - 5)50] = 0.0027 \text{ failures per hour}$$

(2) The mean life is

$$\overline{m} = [\Sigma t_f + (n - f)t]/f$$

$$= [100 + (40 - 5)50]/5 = 370 \text{ hours per failure}$$

$$\overline{m} = 1/\lambda = 1/0.0027$$

$$= 370 \text{ hours per failure}$$

3. The third type of test is also a time-terminated test where the failures are not repaired and replaced but the failure times (of the failed units) are not logged. In this case, an averaging procedure must be used to determine the total test time for the failed units. Two methods can be used.

   a. Assume that each unit failed exactly halfway through the test. The sum of the failure times then is $ft/2$ and the total test time is $ft/2 + (n - f)t$.
   b. Assume that an average number of units completed the test. The total test time, then, is $t[n + (n - f)]/2$.

Example 1.3 Assume that the five failures of Example 1.1 were not repaired and replaced and that no log was kept of their failure times. Find the failure rate and mean life.

Solution:

(1) The failure rate is

$$\lambda = f/[ft/2 + (n - f)t]$$

$$\lambda = 5/[5(50)/2 + (40 - 5)50] = 0.00267 \text{ failures per hour}$$

or

$$\lambda = f/[t(n + (n - f))/2]$$

$$\lambda = 5/[50(40 + (40 - 5))/2] = 0.00267 \text{ failures per hour}$$

(2) The mean life is

$$\overline{m} = [ft/2 + (n - f)t]/f$$

$$\overline{m} = [5(50)/2 + (40 - 5)50]/5 = 375 \text{ hours per failure}$$

or

$$\overline{m} = [t(n + (n - f))/2]/f$$

$$\overline{m} = [50(40 + (40 - 5))/2]/5 = 375 \text{ hours per failure}$$

or

$$\overline{m} = 1/\lambda = 1/0.0267 = 375 \text{ hours per failure}$$

4. The final type of test is a failure terminated test. All units are tested to failure. The total test time, in this case, is the summation of the failure times (Σt).

Example 1.4 Suppose that the 40 units of Example 1.1 were all tested to failure and that the sum of the 40 failure times was 4,000 hr. Find the (1) failure rate and (2) mean life.

Solution:

(1) The failure rate is

$$\lambda = f/\Sigma t = 40/4{,}000 = 0.01 \text{ failure per hour}$$

(2) The mean life is

$$\overline{m} = \Sigma t/f = 4{,}000/40 = 100 \text{ hours per failure}$$

Percentage of Failures

This is sometimes called the ratio failure rate and is just the number of failures per part. It is calculated by dividing the number of parts in the sample (the sample size) into the number of failures. To find the percentage of failures, this value is multiplied by 100. Often a ratio of failures is known from previous studies, from production records, or from quality control records, and this ratio can then be assumed for reliability samples. The equation is

$$r_f = f/n$$

where

$$r_f = \text{ratio of failures} = \text{failures per part (fpp)}$$

$$f = \text{the number of failures in the sample}$$

$$n = \text{the sample size}$$

Example 1.5 A certain production process normally produces 5% scrap. For a sample of 20, how many failures can be expected?

Solution:

$$r_f = f/n$$

$$0.05 = f/20$$

$$f = 0.05(20) = 1 \text{ failure}$$

Hazard Rate

Management usually likes to have some idea of the probability of downtime at any particular instant of time. Since the failure rate is only an average, another measure — the hazard rate — is needed. The hazard rate is defined as the instantaneous failure rate, or the limit of the failure rate as the time interval approaches zero. This must be calculated using the techniques of differential calculus. However, during the mature design phase, after break-in and before wear-out, the failure rate is constant, and so the hazard rate is equal to the failure rate during this phase. The hazard rate during the break-in and wear-out phases usually approximate the constant failure rate, and so this constant failure rate is normally assumed to be the hazard rate during all phases of the system life. (Since the average pattern of failures is often close to constant, this assumption is usually quite valid.)

Terms and Definitions

1. **Reliability.** *Probability* of a device *performing* its function adequately for the period of *time* intended under the operating *conditions* as stated.
2. **Probability.** Chance or likelihood.
3. **Quality.** Degree of conformance to specs.
4. **Standards.** Measure of excellence against which comparisons are made.
5. **Critical defect.** *Will* affect quality.
6. **Major defect.** *May* affect quality.
7. **Minor defect.** *Will not* affect quality (cosmetic defect).
8. **Degradation.** A gradual impairment in ability to perform.
9. **Debugging.** Run until early failures are eliminated (usually at less than rated capacity) — sometimes called derating.
10. **Failure rate (λ).** Equal to number of failures per unit time. ($\lambda = 1/m$; m is the mean life.)
11. **Mean time between failures (MTBF).** Equal to the average time between failures ($1/\lambda$).
12. **Chance defects.** Lack of noticeable pattern. Minute changes due to normal variation of conditions.
13. **Creeping defects.** Gradual reduction of potency or capacity.
14. **Wear-out defects.** The sudden increase in defects after normal wear (or normal product life).
15. **Break-in stage.** The stage in the early life of a product where excess manufacturing defects are detected and corrected.
16. **Operating stage.** The (hopefully) rather long period between break-in and wear-out where failures occur solely by chance and the failure rate is constant.
17. **Wear-out stage.** The last stage in product life, after the operating stage, characterized by a sudden increase in failures (wear-out defects).
18. **Mean time to failure.** Average time to first failure.

Practice Problems

1. Sixty units were tested for a period of 100 hr. At the end of the test period there had been 10 failures, which had immediately been repaired and continued in test. Find the (1) failure rate and (2) mean life.

    Answers: 0.00167; 600

2. Assume that the 10 failures of Problem 1 are not repaired and replaced, and that the failure times are 2, 10, 34, 35, 39, 49, 67, 69, 78, and 96 hr. Find (1) failure rate and (2) mean life.

    Answers: 0.001825; 547.9

3. Assume that the 10 failures of Problem 1 were not repaired and replaced and that no log of failure times was kept. Find (1) failure rate and (2) mean life.

    Answers: 0.00182; 550

4. Suppose that the 60 units of Problem 1 were all tested to failure and that the sum of the failure times for all 60 units was 34,146 hr. Find (1) failure rate and (2) mean life.

    Answers: 0.00176; 569.1

5. A test of 12 units of a certain product was continued until all units failed. The failure times were 5, 12, 22, 34, 35, 45, 51, 65, 67, 78, 81, and 99 hr. Find (1) failure rate and (2) mean life.

    Answers: 0.0202; 49.5

CHAPTER 2 | THE NORMAL DISTRIBUTION

The purpose of this chapter is to explain the fundamentals of the normal curve. This is the most important probability law in all statistics and forms the theoretical foundation for most of the other probability distributions. In reliability, the normal distribution is used to analyze product life during the break-in and wear-out phases, to calculate control limits for mean life, and to determine requirements for spare parts. (The normal distribution is also called "Gaussian.")

Objectives

1. Define and calculate the various measures of central tendency, especially the arithmetic mean and the geometric mean.
2. Define and calculate the various measures of dispersion, with special emphasis on the standard deviation.
3. Understand the differences between a universe (or population) and a sample and between the universe parameters and the sample statistics.
4. Understand the characteristics of the normal curve in relation to mean, mode, median, standard deviation, symmetry, and skewedness.
5. Understand, and be able to use, the Z score and its associated "area under the curve" to solve various types of normal curve problems. Also, understand the use of the "t" distribution.
6. Linearly interpolate to find the correct table value.

2.1 Measures of Central Tendency

A central tendency in statistics is a central or midway value from which other values deviate in some set pattern. It can be thought of as the tendency to be the same. Values tend to group closely around this central value. If the tendency is strong, the values will group very closely to the central value with only a few values at a distance. If the tendency is weak, the values will group loosely about the central value with more values at a larger distance. There are four types of central values: (1) the arithmetic mean, (2) the mode, (3) the median, and (4) the geometric mean.

Arithmetic Mean

The arithmetic mean is the most common and the most useful of all the central values. It is just the sum of all the values divided by the total number of values. There are several ways to calculate the mean, each dependent on the form of the data presentation (how they are organized).

General Formula The general formula for the arithmetic mean is

$$\overline{x} = \frac{\sum_{j=1}^{n} x_i}{n}$$

where

$$\overline{x} = \text{the arithmetic average}$$

$$\Sigma = \text{symbol meaning ``the sum of''}$$

$$x_i = \text{the } i\text{th value}$$

$$n = \text{number of observations}$$

Another form of the above formula is

$$\overline{x} = \frac{x_1 + x_2 + x_3 + \cdots + x_n}{n}$$

where

$$\overline{x} = \text{the arithmetic mean}$$

$$n = \text{the number of observations}$$

$$x_1, x_2, \ldots, x_n = \text{the value of the individual observations}$$

It is often useful to consider an abbreviated form of the above formula that does not contain the formal ith notation (instead the ith notation is just assumed). In the remainder of this text, the abbreviated form of the summation formula will be used with the formal ith notation omitted:

$$\overline{X} = \frac{\Sigma x}{n}$$

Example 2.1 Find the mean of the values 5, 4, 3, 5, 4, 3, 5, 6, 6, 5.

Solution:

$$\overline{X} = \frac{\Sigma x}{n}$$

$$= \frac{5 + 4 + 3 + 5 + 4 + 3 + 5 + 6 + 6 + 5}{10}$$

$$= \frac{46}{10}$$

$$= 4.6$$

Grouped Data It is usually much more useful to summarize a large amount of data by grouping them into cells or classes for more manageable calculations and for better understandability of data presentation. The formula is

$$\overline{X} = \frac{\Sigma(fx)}{n}$$

where

$\overline{X}$ = the arithmetic mean

x = an individual value

n = the total number of observations = Σf

f = the number of observations within each cell (frequency)

Example 2.2 Using the data from Example 2.1, find the mean using the grouped data formula.

Solution:

x	f	fx
3	2	6
4	2	8
5	4	20
6	2	12
	$\Sigma f = n = 10$	$\Sigma(fx) = 46$

$$\overline{X} = \frac{\Sigma(fx)}{n}$$

$$= \frac{46}{10}$$

$$= 4.6$$

Note: This is an example of discrete data where the values are counted in whole numbers only. (Discrete means there can be no division between values — there can never be 2½ defects, for example.) Therefore, in this example, two values measured exactly 3, two values measured exactly 4, etc.

Example 2.3 Find the mean of the following data:

Boundaries	Midpoint (x)	Frequency (f)	fx
55–64	60	5	300
65–74	70	18	1260
75–84	80	30	2400
85–94	90	26	2340
95–104	100	13	1300
		$\Sigma f = n = 92$	$\Sigma(fx) = 7600$

Solution:

$$\overline{x} = \frac{\Sigma(fx)}{n}$$

$$= \frac{7600}{92}$$

$$= 82.61$$

Note: This is an example of continuous data, where the values are measured and can take on any value within the cells. (There can be an infinite division of values between 55 and 65, for example.) In continuous groupings, the values in a cell are assumed to be equal to the midpoint of that cell (five values are equal to 60, etc.). The cell midpoints are the assumed cell means, or averages.

Mode

The mode is that value which has the largest number of readings. In Example 2.1 the mode is equal to 5, and in Example 2.3, the mode is 80. A set of values can be unimodal (one mode), bimodal (two modes), multimodal (many modes), or nonmodal (zero modes). The mode is not used in reliability except when analyzing the normal curve.

Median

The median is that value which is halfway between the lowest value and the highest value. If the set of numbers are in even increments, the median can be determined by adding the lowest and highest values together and dividing by two. Otherwise, the set of values must be organized from lowest to highest, and the median is then found by counting simultaneously from both low and high toward the middle of the set. If there are an odd number of values, the median is the last value reached in the center. If there are an even number of values in the set (or sample), just add the two center values and divide by two. In Example 2.1, the median is 5 [(5 + 5) ÷ 2]. The median is not used in reliability except as a tool for analyzing the normal curve.

Geometric Mean

The geometric mean is determined by successively multiplying the set (sample) of values and then taking the nth root of the result. The formula is

$$GM = \sqrt[n]{\Pi x}$$

where

GM = the geometric mean

n = the total number of observations

x = an individual value

Π = symbol meaning successive multiplication

Example 2.4 Using the data from Example 2.1, find the geometric mean.

Solution:

$$GM = \sqrt[n]{\Pi x}$$

$$= \sqrt[10]{3 \times 3 \times 4 \times 4 \times 5 \times 5 \times 5 \times 5 \times 6 \times 6}$$

$$= \sqrt[10]{3,240,000}$$

$$= 4.478$$

In reliability, the geometric mean is used to compute equivalent component reliability and is also useful in the analysis and prediction of redundant systems.

2.2 Measures of Dispersion

Dispersion is the tendency to be different or the variability of the values. This is the principle that causes the values to vary about the mean, or central value. If this tendency is strong, the spread of values from the central tendency will be greater and the average distance from the central value will be larger. There are three types of variability, or dispersion, measures: (1) range, (2) standard deviation, and (3) variance.

Range

The range is the difference between the highest and the lowest values. The formula is

$$R = x_h - x_l$$

where

$$R = \text{the range}$$

$$x_h = \text{the highest value}$$

$$x_l = \text{the lowest value}$$

Example 2.5 Find the range of data in Example 2.1.

Solution:

$$R = x_h - x_l$$

$$= 6 - 3$$

$$= 3$$

Example 2.6 Find the range of the data of Example 2.3.

Solution:

$$R = x_h - x_l$$

$$= 105 - 55$$

$$= 50$$

In this example, the lowest value is assumed to be the lower boundary of the lowest cell (55) and the highest value is assumed to be the highest boundary of the highest cell (105), although, in actual fact, these boundaries could have zero frequencies. If the real data are known (available), then the actual lowest and highest values would be used. The range is an inefficient means of measuring dispersion. Although used extensively in quality control to approximate the standard deviation, it has almost no usefulness in reliability calculations.

Standard Deviation

The standard deviation is defined as the square root of the sum of the squares of the differences between the individual values and the mean divided by the number of values (or measurements). The standard deviation is a measure of the average distance that the values deviate from the central value, or mean. Assume a mean of 6 and a standard deviation of 2. A number line can then be established with the mean at 6, as follows.

$$-\infty\ \overline{1\ 2\ 3\ 4\ 5\ 6\ 7\ 8\ 9\ 10}\ +\infty$$
$$\overline{x}$$

From 4 to 6 is one standard deviation, from 6 to 8 is another, from 8 to 10 is another, from 1 to 3 is another, and so forth. Two standard deviations would be equal to 4 units on the number line such as from 1 to 4, or from 5 to 9, or from 4 to 8, etc. The standard deviation is the most useful and most used measure of dispersion. There are three main formulas for the standard deviation: (1) the basic formula, (2) the sum of squares formula, and (3) the grouped data formula. The grouped data formula can be used for both discrete and continuous data.

Basic Formula The basic formula is

$$s = \sqrt{\frac{\Sigma(x - \overline{x})^2}{n - 1}}$$

where

s = the standard deviation of a sample
(the formula is slightly different for the standard deviation of the universe)

x = an individual value

$\overline{x}$ = the arithmetic mean

n = the number of values
(or measurements or observations)

Σ = "the sum of"

Example 2.7 Find the standard deviation of the data of Example 2.1.

Solution:

x	$\overline{x}$	$x - \overline{x}$	$(x - \overline{x})^2$
3	4.6	−1.6	2.56
3	4.6	−1.6	2.56
4	4.6	−0.6	0.36
4	4.6	−0.6	0.36
5	4.6	0.4	0.16
5	4.6	0.4	0.16
5	4.6	0.4	0.16
5	4.6	0.4	0.16
6	4.6	1.4	1.96
6	4.6	1.4	1.96
$\Sigma x = 46$		$\Sigma(x - \overline{x}) = 0.0$	$\Sigma(x - \overline{x})^2 = 10.40$

$$\overline{x} = \frac{\Sigma x}{n} = \frac{46}{10} = 4.6$$

$$s = \sqrt{\frac{\Sigma(x - \overline{x})^2}{n - 1}}$$

$$= \sqrt{\frac{10.4}{10 - 1}}$$

$$= 1.07$$

Note: The sum of $x - \overline{x}$ is equal to zero and therefore cannot be used in further calculations (the variation would always be equal to zero). This is the reason that $(x - \overline{x})^2$ is used to find the variation.

The "Sum of the Squares" Formula Using basic algebra, a formula for the standard deviation can be derived that does not require the calculation of the mean ($\overline{x}$).

$$(x - \overline{x})^2 = x^2 - 2x\overline{x} + \overline{x}^2$$

summing

$$\Sigma x^2 - 2\Sigma x\overline{x} + n\overline{x}^2$$

substituting

$$\Sigma x^2 - 2\Sigma x\frac{\Sigma x}{n} + n\left(\frac{\Sigma x}{n}\right)^2$$

$$= \Sigma x^2 - \frac{2(\Sigma x)^2}{n} + \frac{(\Sigma x)^2}{n}$$

$$= \Sigma x^2 - \frac{(\Sigma x)^2}{n}$$

This is called the "sum of squares" and is extremely useful in advanced analysis techniques such as design of experiments and analysis of variance (ANOVA).

Substituting into the basic formula yields

$$s = \sqrt{\frac{\Sigma x^2 - \frac{(\Sigma x)^2}{n}}{n - 1}}$$

This method is ideally suited for a calculator or computer. In a computer only three totals would have to be maintained for incoming data: (1) Σx (just add each new value to the previous sum), (2) Σx^2 (just add the square of each new value to the previous sum), and (3) n (just add 1 for each new value entered). In this way a running value for the mean and the standard deviation could be computed at any time (say daily or weekly).

Example 2.8 Compute the standard deviation of Example 2.1 using the "sum of squares" formula.

Solution:

x	x^2
3	9
3	9
4	16
4	16
5	25
5	25
5	25
5	25
6	36
6	36
$\Sigma x = 46$	$\Sigma x^2 = 222$

$$s = \sqrt{\frac{\Sigma x^2 - \dfrac{(\Sigma x)^2}{n}}{n - 1}}$$

$$= \sqrt{\frac{222 - \dfrac{(46)^2}{10}}{10 - 1}}$$

$$= 1.07$$

Note: The mean can still be computed with this method by using Σx and n:

$$\overline{x} = \frac{\Sigma x}{n} = \frac{46}{10} = 4.6$$

Grouped Data Large amounts of data need to be grouped or organized into some kind of orderly arrangement in order for the data to be meaningful and useful. The best method is to group the data into even frequencies called cells or classes. The formula is

$$s = \sqrt{\frac{\Sigma(fx^2) - \dfrac{(\Sigma fx)^2}{n}}{n - 1}}$$

There are basically two methods of grouping data: (1) group the individual values in order from lowest to highest (used mostly for discrete data) and (2) group into cells of even intervals and organize from lowest to highest (used mostly for continuous data).

Example 2.9 Using the discrete data of Example 2.1, calculate the standard deviation using the grouped data method.

Solution:

x	f	fx	x^2	fx^2
3	2	6	9	18
4	2	8	16	32
5	4	20	25	100
6	2	12	36	72
$\Sigma f = n = 10$		$\Sigma(fx) = 46$		$\Sigma(fx^2) = 222$

$$s = \sqrt{\frac{\Sigma(fx^2) - \frac{(\Sigma fx)^2}{n}}{n - 1}}$$

$$= \sqrt{\frac{222 - \frac{(46)^2}{10}}{10 - 1}}$$

$$= 1.07$$

Note: The mean can also be calculated from this method:

$$\bar{x} = \frac{\Sigma(fx)}{n} = \frac{46}{10} = 4.6$$

Example 2.10 Using the continuous data of Example 2.3, calculate the standard deviation using the grouped data method.

Solution:

Midpoint (x)	f	fx	x^2	(fx^2)
60	5	300	3600	18,000
70	18	1260	4900	88,200
80	30	2400	6400	192,000
90	26	2340	8100	210,600
100	13	1300	10000	130,000
$\Sigma f = n = 92$		$\Sigma(fx) = 7600$		$\Sigma(fx^2) = 638,800$

$$s = \sqrt{\frac{\Sigma(fx^2) - \frac{(\Sigma fx)^2}{n}}{n - 1}}$$

$$= \sqrt{\frac{638,800 - \frac{(7600)^2}{92}}{92 - 1}}$$

$$= 10.98$$

Note: The mean can be calculated by

$$\overline{x} = \frac{\Sigma(\,fx)}{n} = \frac{7600}{92} = 82.61$$

Variance

The variance is just the square of the standard deviation. It can be computed by using any of the standard deviation formulas and omitting the radical (do not take the square root) or by just squaring the standard deviation. In the two examples above the variance is 1.145 (1.07^2) and 120.56 (10.98^2). The symbol for the variance is s^2.

2.3 Concept of a Universe and a Sample

The collection of all possible values is known as a universe or population. For example, a list of IQ's of all American males would be a population, or universe. Suppose it were desired to know the average IQ of all American males. One method of achieving this goal would be to measure the IQ of each American male, sum the values, and divide by the total number of American males. Obviously this would be a tedious, expensive, and, probably, impossible task. These disadvantages can be largely overcome by means of sampling. A sample (say 100 American males) can be measured and related directly to the universe from which it was taken. This sample must be drawn randomly and must be homogeneous; it must be representative of the universe from which it was drawn. Randomality means that each item (person) in the universe has an equal chance of being chosen for the sample. Homogeneity means that all groups within the universe are represented in the sample in the same proportion in which they occur in the universe. A sample that is truly random and homogeneous is termed an unbiased sample, while one that is not random nor homogeneous is considered to be biased.

Frequency Groupings

There are four main methods used to illustrate a distribution, most of them related to the frequency (the number of times that a particular value occurs in the set). One method is a simple list of the values of the set organized from lowest to highest along with their associated frequencies (in a separate column). In Example 2.3, the list of the midpoints, along with the list of their frequencies (only these two columns) constitutes this type of frequency distribution. If these frequencies are arranged in bars on a chart (Fig. 2.1), the resultant chart is called a histogram (or bar chart). If they are located as points on a graph with connecting lines (Fig. 2.2), the resultant chart is

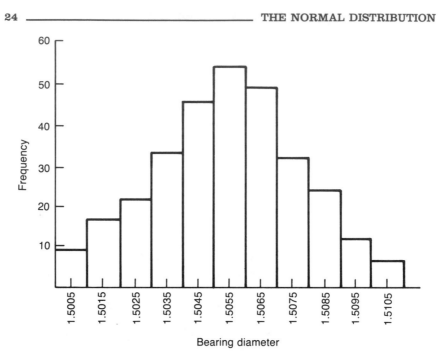

Fig. 2.1. Histogram of bearing diameters.

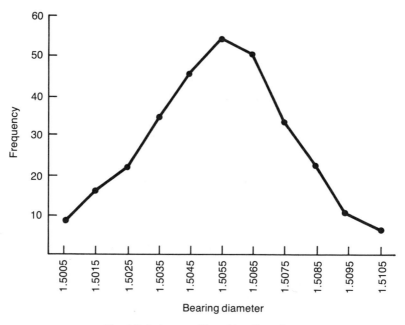

Fig. 2.2. Polygon of bearing diameters.

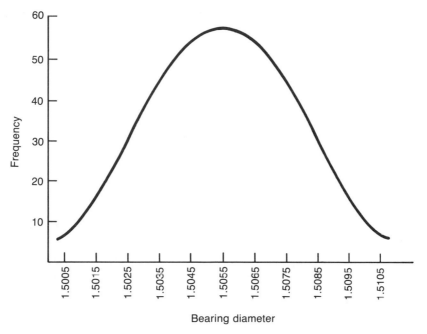

Fig. 2.3. Frequency distribution curve.

called a frequency polygon. And, finally, if the graph represents a continuous distribution (with an infinite number of points, theoretically, between each value), the resultant chart is called a frequency distribution curve (Fig. 2.3). A frequency distribution curve can be almost any shape, but only a few are of statistical importance.

Parameters versus Statistics

The measures of the characteristics of the universe are called parameters, while the measures of the characteristics of the sample are called statistics. Ideally, statistics should closely resemble and approximate the parameters they represent. In order to differentiate between parameters and statistics, different symbols are used. Thus, the symbol for a sample mean (statistic) is $\bar{x}$, while the symbol for a universe mean (parameter) is the Greek letter μ. Similarly, the statistical symbol for the standard deviation of a sample is s, while the parameter symbol for the standard deviation of a universe is the Greek letter σ. Unfortunately, these symbols are not standardized in the literature. Different authors frequently use different symbols such as $\bar{x}$ for a sample mean, $\bar{x}'$ for the universe mean, σ for the sample standard deviation, and σ' for the standard deviation of a universe. However, these symbols ($\bar{x}$, μ, σ, and s) are probably the most used in statistical literature. It should be noted that while the formulas for the means of a sample and for a universe are identical ($\bar{x} = \mu$), the formulas for the two standard deviations are not. The formula for the universe standard deviation is

$$\sigma = \sqrt{\frac{\Sigma (x - \mu)^2}{n}}$$

while the formula for the sample standard deviation is

$$s = \sqrt{\frac{\Sigma (x - \bar{x})^2}{n - 1}}$$

Since $\mu = \bar{x}$ (assuming the universe and the sample are both normally distributed and that the sample is unbiased), the only real difference between the two formulas is that one is divided by n, while the other (the sample) is divided by $n - 1$. The reason for this is to offset the normal bias of small sample sizes. When the sample size gets large (100 or more), the difference between the two formulas is nil (dividing by 100 gives almost identical results to dividing by 99, while dividing by 10 will give quite different results than dividing by 9).

Why Sample?

As has already been shown, it is often impossible to measure all the items in a distribution. There are five reasons why a sample is usually more desirable than 100% inspection. First, the distribution may be so large or unusual that it is impossible to measure all its units. For instance, it would be impossible to measure the height of all human beings (and record the values in a distribution), since, by the time the job was done, there would be so many new humans that the job could never be finished. Second, even if it were possible to actually measure and record the height values, the job would be so costly that it would not be worth it — no benefit could possibly offset the prohibitive cost involved. A third reason for sampling is when the measurement destroys the product. Under this condition, 100% inspection would leave no product to sell. A fourth reason for sampling is that some products might be so dangerous, such as radioactive or molten materials, that 100% inspection would subject the measurers to too much danger. Finally, there are psychological factors that suggest that sampling may actually be more accurate than 100% inspection; 100% inspection leads to boredom and fatigue that usually result in more errors and less reliable information than does sampling.

Objective of Sampling

The reason for sampling is to learn something, or infer something, about the universe from which the sample was drawn. If the sample were properly drawn, at random and without bias, the information learned from the sample can be extrapolated to the universe. The universe should then be similar in its characteristics to the sample characteristics.

2.4 Characteristics of the Normal Curve

The normal curve is quite distinctive in shape. It is symmetrical, unimodal, and bell-shaped with the mean, mode, and median being equal. A symmetrical object, when folded along its center line (the mean in this case), will have the outline of one side exactly cover the outline of the other side. Unimodal means there is only one high spot in the curve. The series of graphs in Fig. 2.4 illustrates these relationships. Figure 2.4a shows a normal, symmetrical, bell-shaped curve. Figure 2.4b shows a distribution that is skewed to the right. (Skewedness always refers to the direction of the long tail.) This curve is also positively skewed (positive to the right; negative to the left). Figure 2.4c shows a negatively skewed curve (skewed to the left). Figures 2.4d and 2.4e show the leptokurtic-type (from *kurtosis* meaning peakedness) distribution, with its high peak and narrow dispersion (strong central tendency) and the platykurtic-type distribution, with its low central peak and wide distribution (weak central tendency).

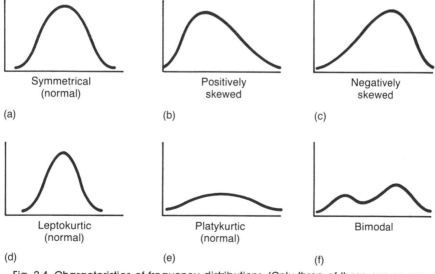

Fig. 2.4. Characteristics of frequency distributions. (Only three of these curves are normal curves.)

Leptokurtic curves have small standard deviations, while platykurtic standard deviations are quite large in comparison. Figure 2.4f shows a bimodal distribution with two humps. Data that were used to construct this curve probably came from two separate universes (such as two different machines).

Relationship Among Mean, Mode, and Median

Figure 2.5 gives further comparisons of the relationships of mean, mode, and median in various types of curves. In the normal, symmetrical, curve (Fig. 2.5a), the mean = the mode = the median. If the curve is skewed to the right (Fig. 2.5b), the mean is displaced to the right, and the median provides a more representative value for the central tendency. Similarly, in left, or negatively, skewed distributions (Fig. 2.5c), the mean is displaced to the left and the median again provides a better representation of the central value.

Construction and Analysis of the Normal Curve

The first step in constructing a normal curve is to draw a number line horizontally from left to right (similar to the x axis of the cartesian coordinate system). This line is assumed to contain all possible values (numbers) stretching from $-\infty$ on the left to $+\infty$ on the right:

$-\infty$ —— $+\infty$

0

Next a line is assumed perpendicular to this number line (identical to the y axis) and marked off from 0 to $+\infty$. This perpendicular line always locates the number of values while the horizontal line locates the actual, individual, unit values. For instance, if it is desired to graph the IQ's of all American males, the horizontal scale would locate each individual IQ, while the perpendicular scale would show the number (frequency) of American males at each IQ level. If the number of American males with a particular IQ were known (say 100,000 have an IQ of 105), a point on the graph would be found by reading

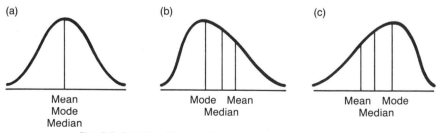

(a)

(b)

(c)

Mean	Mode	Mean	Mean	Mode
Mode	Median		Median	
Median				

Fig. 2.5. Relationship among mean, mode, and median.

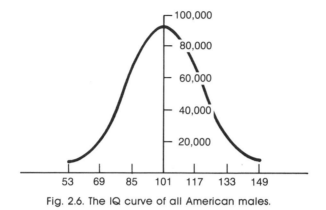

Fig. 2.6. The IQ curve of all American males.

across on the number line to 105 and up on the frequency line (perpendicular line) to 100,000. Similarly, all points on the graph would be located and a curve plotted as in Fig. 2.6. Once plotted, the number of American males having a particular IQ could be found from the curve by locating the IQ on the horizontal number line, then moving up (perpendicular) from the line to the curve, and then moving to the left or right (horizontally) to the frequency line and reading the value (frequency) at that point.

Relationships of Mean and Standard Deviation

It has been found that the mean IQ of American males is 101 with a standard deviation of 16. The mean of 101 locates the IQ curve on the number line. There can be an infinite number of curves in the number line (from $-\infty$ to $+\infty$) with the same standard deviation (Fig. 2.7). The standard deviation, on the other hand, defines the shape of the curve; it has no effect on its location. A small standard deviation makes the curve narrow and high peaked, while a large standard deviation causes the curve to be shallow and long (Fig. 2.8). There can also be an infinite number of standard deviations with identical means (an infinite number with a mean of 101, an infinite number with a mean of 100, etc.).

z Scores

The number of standard deviations from the mean is a very important value in the analysis of the normal curve. Since the standard deviation is just the distance of the average dispersion on the number line (see Section 2.2), the number of these distances (the number of standard deviations) from the mean is found by the formula

$$z = \frac{x - \mu}{\sigma}$$

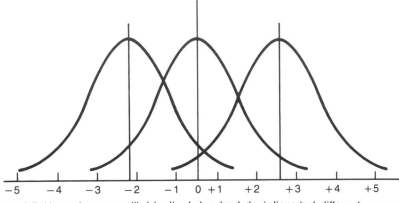

Fig. 2.7. Normal curves with identical standard deviations but different means.

where

z = the standardized normal value

x = an individual observation or measurement

μ = the arithmetic mean of the universe

σ = the standard deviation of the universe

z, then, is just the number of standard deviations from the mean.

Area Under the Curve

Equations can be developed for any curve and the normal is no exception. Thus, the area under this curve can be found by the techniques of integral calculus. This area can be directly related to the percentage of items in the distribution. Therefore, the area under the curve from $-\infty$ to $+\infty$ represents 100% of all items in the distribution.

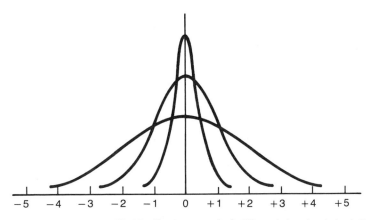

Fig. 2.8. Normal curves with identical means but different standard deviations.

The area under the curve from $-\infty$ to the mean (and from the mean to $+\infty$) represents 50% of all the items in the distribution. This can be verified by actually calculating the total area under a curve (say the IQ curve) and dividing it into the area under the curve from $-\infty$ to the mean. Similarly, any area, from any value to any other value (say from IQ 90 to IQ 105) can be calculated and the total area divided into it to get the percentage of items in the distribution between the two values. Then, if the total number of items in the distribution is known (total number of American males), this total can be multiplied by the percentage between the two values to get the actual number of items that lie between these two values (actual number of American males with IQ's between 90 and 105, for instance). This process is quite tedious and time consuming, therefore, a table of these percentages (listed as ratios rather than percentages) has been calculated and is given in the Appendix (Table A1).

Standardized Normal Curve

Table A1 has been computed by integrating the area under one particular curve, called the standardized normal curve. This curve has a mean of 0, a standard deviation of 1, and an area under the curve of 1.00 (Fig. 2.9). Thus, many of the terms of the basic formula are simplified for calculation purposes. All terms that are multiplied or divided by the mean reduce to zero and are eliminated, while terms that are multiplied or divided by the standard deviation remain the same. (Any term that is multiplied or divided by 1 does not change its value.) This modified equation is

$$f(z) = y = \frac{1}{\sqrt{2\pi}} \, e^{-z^2/2}$$

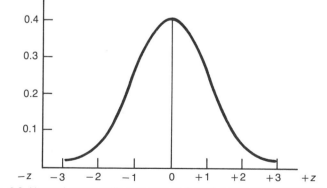

Fig. 2.9. Normal curve with a mean of 0 and a standard deviation of 1.

In this curve, the area under the curve from one value to another also represents the percentage of items between these two values. Since the total area of this curve is equal to 1.00, dividing by the total area to get the percentage is unnecessary.

All normal curves are symmetrical, therefore the percentage area under the curve from the mean to any particular z (number of standard deviations) is always the same for all normal curves. For example, even though the standard deviations (and the means) may be different, the percentage areas from the mean to one, two, or three standard deviations (z = 1, 2, or 3) are always the same. This also holds true, of course, for any value in between (z = 1.5, 1.8, 2.9, 3.5, etc.). Figure 2.10 illustrates these concepts showing the percentage areas for one, two, and three standard deviations.

Using the Table

To use the table of areas under the curve, first calculate the mean, the standard deviation, and the z score. Then enter the table at the proper z score to get the area (from $-\infty$ to x). The following examples will be used to illustrate the use of the table.

Example 2.11 For a mean of 20 and a standard deviation of 2, find the area under the curve at x = 16 or below.

Solution:

$$z = \frac{x - \mu}{\sigma} = \frac{16 - 20}{2} = \frac{-4}{2} = -2.00$$

$$p(z \leqslant -2.00) = 0.0228 \text{ or } 2.28\%$$

Enter the table under the column headed z. Find -2.00 in this column. Move horizontally on this line to find the value under the column headed 0.00. This value, then, is the percentage area under the curve from $-\infty$ to 16 (percentage of items at 16 or less).

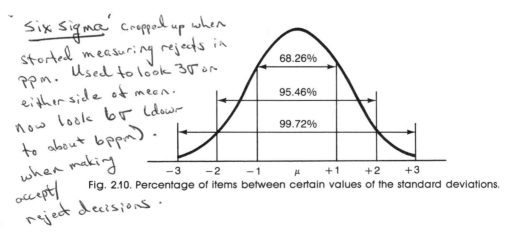

Six Sigma' cropped up when started measuring rejects in ppm. Used to look 3σ on either side of mean. Now look 6σ (down to about 6ppm). when making accept/reject decisions.

68.26%

95.46%

99.72%

−3   −2   −1   μ   +1   +2   +3

Fig. 2.10. Percentage of items between certain values of the standard deviations.

Example 2.12 For a mean of 20 and a standard deviation of 3, find the area at 22 or less.

Solution:

$$z = \frac{22 - 20}{3} = \frac{2}{3} = 0.667$$

$$p(z \leqslant +0.667) = 0.7475 \text{ or } 74.75\%$$

Enter the table under the z column at $+0.6$, move to the columns headed 0.06 and 0.07. The percentage area at 0.66 is 0.7454, while the percentage area for $z = 0.67$ is 0.7486. Interpolating (add 70% of the difference to 0.7454) gives 0.7475.

Example 2.13 For a percentage of 85%, find the z score (number of standard deviations from the mean).

Solution:

When a percentage is given, the process is reversed. Enter the body of the table and find 85% (0.8500), which lies between 0.8485 and 0.8508. At 0.8485, $z = 1.03$ and at 0.8508, $z = 1.04$. Interpolating gives a z of 1.0365 at 85%.

2.5 Applications of the Normal Curve

The normal curve can be used to determine a percentage of items below a certain value, above a certain value, or between two values. It can also be used inversely to determine a desired value or mean from a given percentage of items.

To Find the Percentage of Items Below a Given Value

The steps used to find the percentage of items below a given value are:

1. Draw the normal curve.
2. Locate all given and calculated values on the curve. (This step provides a picture of the problem and assists in directing the solution.)
3. Calculate z.
4. Obtain the answer from table.

Example 2.14 The mean weight of a product is 0.302 kg with a standard deviation of 0.025 kg. Find the percentage of product below 0.280 kg.

Solution:

$$z = \frac{x_i - \mu}{\sigma} = \frac{0.280 - 0.302}{0.025} = -\frac{0.022}{0.025} = -0.88$$

$$p(z \leqslant -0.88) = 0.1894 \text{ or } 18.94\%$$

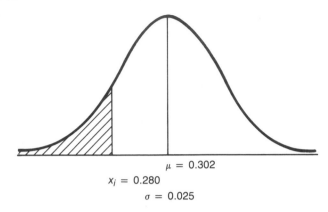

$$\mu = 0.302$$
$$x_i = 0.280$$
$$\sigma = 0.025$$

Note: 18.94% of the items are below 0.280 kg (in a large sample) *or* the chance of any one item being below 0.280 is 18.94% *or* out of 10,000 boxes, 1894 will weigh less than 0.280 kg while 10,000 − 1894 or 8106 will weigh more.

To Find Percentage of Items Above a Certain Value

The steps used to find the percentage of items above a certain value are:

1. Draw the normal curve.
2. Locate all values on the curve.
3. Calculate z.
4. Obtain value from table.
5. Subtract table value from 1.00.

Example 2.15. From the data of Example 2.14, find the percentage of product above 0.376 kg.

Solution:

$$z = \frac{x_i - \mu}{\sigma} = \frac{0.376 - 0.302}{0.025} = \frac{0.074}{0.025} = 2.96$$

$$p(z > 2.96) = 1 - p(z \leqslant 2.96) = 1 - 0.9985 = 0.0015 \text{ or } 0.15\%$$

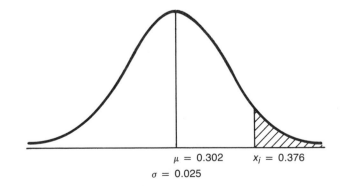

$$\mu = 0.302 \qquad x_i = 0.376$$
$$\sigma = 0.025$$

To Find Percentage of Items Between Two Values

The steps used to find the percentage of items between two values:

1. Draw the curve.
2. Locate all values on the curve.
3. Calculate z for each given value.
4. Find table value for each z.
5. Subtract smaller table value from larger.

Example 2.16 From the data in Example 2.14, find the percentage of product between 0.280 and 0.376 kg.

Solution:

$$z_1 \text{ (from Example 2.14)} = -0.88$$

$$p(z_1 < -0.88) = 0.1894$$

$$z_2 \text{ (from Example 2.15)} = 2.96$$

$$p(z_2 < 2.96) = 0.9985$$

Therefore,

$$p(z_2) - p(z_1) = 0.9985 - 0.1894 = 0.8091 \text{ or } 80.91\%$$

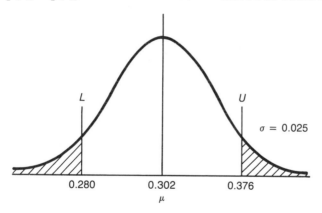

The Inverse Use of the Normal

The steps for the inverse use of the normal are:

1. Draw the curve.
2. Locate all values on the curve.
3. Find z score from the given percentage.
   Note: If the percentage is above the desired value (or specification), subtract it from 1.00 before finding the z score.
4. Calculate the mean or other value from the z formula.

Example 2.17 Using the data from Example 2.14, find what the mean (or machine setting) would have to be in order to ensure that no more than 5% of the items weigh below 0.280 kg.

Solution:
From Table A1, $z = -1.645$ at 0.0500. Therefore,

$$z = -1.645 = \frac{0.280 - \mu}{0.025}$$

$$\mu = 1.645\,(0.025) + 0.280$$

$$\mu = 0.321 \text{ kg}$$

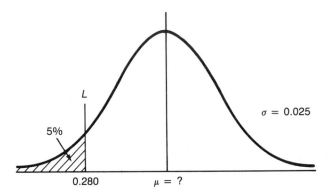

Example 2.18 How does this mean found in Example 2.17 (0.321 kg) affect the percentage of items above the upper specification limit of 0.376?

Solution:

$$z = \frac{0.376 - 0.321}{0.025} = \frac{0.055}{0.025} = 2.20$$

$$p(z > 2.20) = 1 - p(z < 2.20) = 1 - 0.9861 = 0.0139 \text{ or } 1.39\%$$

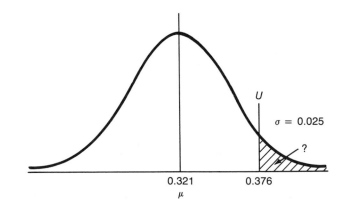

Example 2.19 What would the upper specification limit have to be if the mean found in Example 2.17 must remain the same (0.302 kg) and 5% of the items can be above the upper specification (standard deviation = 0.025)?

Solution:

From Table A1, $z_{(1-0.05)} = 1.645$

$$z_{0.95} = 1.645 = \frac{x - 0.302}{0.025}$$

$$x = 1.645\,(0.025) + 0.302$$

$$x = 0.343 \text{ kg}$$

The *t* Distribution

In the previous six examples (2.14 through 2.19), the means and standard deviations were assumed to be universe parameters that were derived from the entire population of product. For a long-term project, or on-going production, where a wealth of information is available about past performance (say the previous year's production of cereal boxes), this assumption is usually quite valid. (The calculated parameters from this much data can be assumed to equal the real universe parameters.) If these kinds of data are not available, the universe parameters must be estimated from sample data ($\bar{x}$ and s instead of μ and σ) and a new distribution, the *t* distribution, must be used instead of the normal distribution (sometimes universe values, mean and standard deviation, are assumed by management or engineering or both for analysis purposes). Such is usually the case in reliability testing.

The *t* distribution is actually the normal adjusted to overcome inherent errors due to small sample sizes and to provide a safety factor for the estimates. Thus, as the sample size increases, the *t* distribution approaches the normal (at about $n = 30$). The *t* distribution is used in the same way as the normal except that the t_α values (area under the curve or percentage of items) depend on the size of the sample and the table must be entered at $n - 1$ as well as at t. The formula for the *t* score is

$$t_{\alpha, n-1} = \frac{x - \bar{x}}{s}$$

where

α = area under the curve below the critical value, x

t = the number of standard deviations from the mean of a *t* distribution

x = desired value or measurement

$$\overline{x} = \text{mean of the sample}$$

$$s = \text{standard deviation of the sample}$$

As with the normal curve, the t distribution can be used to find the percentage of items below or above a certain value and between two values, and a desired mean or value from a given percentage. To use the t table (Table A4), enter the table at $n - 1$, move to the right to find the t value (it may be necessary to interpolate) and read the P value (probability value or α) from the top of the table (if t is positive) or from the bottom of the table (if t is negative). To find a t value from a given sample size and probability value (inverse use of the table), enter the table at the desired probability value, move down (or up) the column until the desired degree of freedom line is reached ($n - 1$), and read the t value from the table (it may be necessary to interpolate). If the P value (α) is less than 0.50, the t value is negative.

Example 2.20 Using the data from Example 2.14, except that the sample size is now 10, find the percentage of product below 0.280 kg. The mean is 0.302 and the standard deviation is 0.025. Note that the degrees of freedom ($n - 1$) is $10 - 1 = 9$.

Solution:

$$t_{\alpha,10-1} = \frac{x - \overline{x}}{s} = \frac{0.280 - 0.302}{0.025} = -0.88$$

$$P(t_{\alpha,10-1} \leqslant -0.88) = 0.2009 \text{ or } 20.09\% \text{ (Table A4)}$$

Note the difference between the above answer and the answer obtained using the normal distribution (18.94%) in Example 2.14.

Example 2.21 Using the data from Example 2.14, find the percentage of items above 0.376 kg when the mean is 0.302 kg, the standard deviation is 0.025, and the sample size is 10 (compare to Example 2.15).

Solution:

$$t_{\alpha,10-1} = \frac{x - \overline{x}}{s} = \frac{0.376 - 0.302}{0.025} = 2.96$$

$$P(t_{\alpha,10-1} > 2.96) = 1 - P(t_{\alpha,10-1} \leq 2.96) = 1 - 0.9985 = 0.15\% \text{ (Table A4)}$$

Note the difference between the above answer and the answer obtained from Example 2.15 (0.15%).

Example 2.22 Using the data from Example 2.14, find the mean machine setting if no more than 5% of the items are to be below 0.280 kg.

Solution:

$$t_{0.05,10-1} = -1.833 \text{ (Table A4)}$$

$$t_{0.05,10-1} = -1.833 = \frac{0.280 - \bar{x}}{0.025}$$

$$\bar{x} = 0.326 \text{ kg}$$

Compare to 0.321 kg in Example 2.17.

Statistical Accuracy

Owing to the nature of day-to-day living, most of us are conditioned to consider a difference of one-tenth of one percent (0.001) or less as insignificant; thus, rounding to two or three decimal points impresses us as being more than adequate. However, in statistics, this is seldom acceptable. It is important, at this point, to emphasize the need for, at least, four-significant-decimal-place accuracy in statistics. This means that all final answers to problems must be rounded to four decimal places, even though the elements of the problem were presented in less than four decimal places. Although this may appear to be nit-picking, an example should suffice to justify such a rule.

Suppose that there are two methods of producing the same product. Management wishes to know if the two methods are comparable; if they can both produce reliable product. A study shows that the reliability of the two methods are 0.9900 and 0.9947, respectively. If the figures are rounded to two places, and reported as such, it would appear to management that the two systems are comparable and can be expected to produce product at the same reliability level (the number of failures will be the same). However, a subsequent analysis of failed units shows that the first system has produced twice as many failed units as has the second method. Management wants to know why. If the data had been originally reported to four-decimal place accuracy, the reason would have been obvious. Out of 10,000 produced units, the first method can be expected to produce 100 failures (10,000 − 9900), while the second method will produce only 53 failures (10,000 − 9947) or one-half as much.

2.6 Linear Interpolation

The first step in linear interpolation is to set up the known and unknown values. A ratio equation is then set up and solved for the unknown value, say, *c*:

$$\frac{a}{b} = \frac{c}{d}$$

$$c = \frac{ad}{b}$$

a, b, c, and d represent the differences between the numbers shown in the tables:

Example 2.23 Needed: Z for $P = 0.20$.

Solution:

$$a = 0.1977 - 0.2000 = -0.0023$$

$$b = 0.1977 - 0.2005 = -0.0028$$

$$c = \text{Unknown}$$

$$d = -0.85 - (-0.84) = -0.01$$

$$c = \frac{ad}{b} = \frac{-0.0023(-0.01)}{-0.0028} = -0.00082$$

Now, Z = unlisted value = $-0.85 - (-0.00082) = -0.8418$

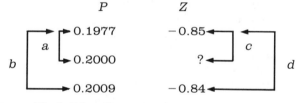

Example 2.24 Find Z for $P = 0.85$.

Solution:

$$a = 0.8485 - 0.8500 = -0.0015$$

$$b = 0.8485 - 0.8508 = -0.0023$$

$$c = \text{Unknown}$$

$$d = 1.03 - 1.04 = -0.01$$

$$c = \frac{ad}{b} = \frac{-0.0015\,(-0.01)}{-0.0023} = -0.00652$$

Now, $Z = 1.03 - (-0.00652) = 1.03652$

Example 2.25 Find P for $Z = 1.264$ ($Z_{1.264} = ?$).

Solution:

$$a = 1.26 - 1.264 = -0.004$$

$$b = 1.26 - 1.27 = -0.010$$

$$c = \text{Unknown}$$

$$d = 0.8962 - 0.8980 = -0.0018$$

$$c = \frac{ad}{b} = \frac{-0.004\,(-0.0018)}{-0.010} = -0.00072$$

Now, $P = 0.8962 - (-0.00072) = 0.89692\ (89.7\%)$

Terms and Definitions

1. **Central tendency.** Tendency to be the same.
2. **Mean.** Arithmetic average $= \bar{x}$ — the most important measure of the central tendency.
3. **Mode.** Value with the greatest frequency — a measure of central tendency.
4. **Median.** Value halfway between the highest and lowest values — a measure of central tendency.
5. **Dispersion.** Tendency to be different.
6. **Range.** Difference between the highest and lowest values — a measure of dispersion.
7. **Standard deviation.** s = average dispersion of the values of a set — the most important measure of dispersion.
8. **Grouped data.** Values grouped together in cells for easier analysis.
9. **Normal curve.** The symmetrical bell-shaped curve formed when large amounts of data are graphed.
10. **Unimodal.** One mode or one hump in the curve.
11. **Universe (population).** Collection of all possible values.
12. **Sample.** A small group of values chosen at random from the universe.
13. **Randomality.** Equal chance of being chosen.
14. **Parameter.** Characteristic of the universe.
15. **Statistic.** Characteristic of the sample.
16. **Symmetrical.** When folded along the center line, the outline of one side exactly covers the outline of the other.
17. **z score.** The number of standard deviations from the mean.
18. **Percentage area under the curve.** Area between two values divided by total area under the curve. Also found in the body of the "area under the curve" table.

Practice Problems

1. What is the mean and standard deviation of: 23.9, 22.4, 25.6, 19.1, 19.4, 17.9, 20.0, 23.6, and 18.8?

   Answers. 21.188; 2.729

2. What is the mean and standard deviation of the following frequency distribution?

Cell Midpoint	Frequency
40	2
48	4
56	5
64	11
72	19
80	31
88	43
96	36
104	26
112	12
120	8
128	2
136	1

   Answers. 88.72; 17.186

3. If the mean weight of a company's product is 21.2 lb and the standard deviation is 1.7 lb, find: (a) the percentage below 18 lb; (b) the percentage above 22.5 lb; and (c) the percentage between 18 and 22 lb.

   Answers: 0.0299; 0.2222; 0.6511

4. A manufacturer wishes to have 2% of product to be below 9.5 in. in length. If the standard deviation is 0.42 in., what mean length is required?

   Answer: 10.3627 in.

5. The 100-hr reliabilities of three components of a radio receiver are 0.97, 0.98, and 0.95. What is the geometric mean?

   Answer: 0.9666

6. Ten light bulbs were tested to failure. The hours to failure were 221, 385, 501, 630, 672, 682, 700, 715, 875, 900. What is the mean time between failures and the standard deviation?

   Answers: 628.1; 209.1

7. The mean life and standard deviation of a certain vacuum tube were found to be 2000 hr and 100 hr, respectively. If the data are normally distributed, find (a) the percentage of tubes expected to

exceed 2100 hr; (b) the percentage expected to last for 1900 hr or less; and (c) the percentage between 1900 and 2100 hours.

Answers: 15.87; 15.87; 68.26

8. A manufacturer of breakfast cereal wishes to keep his product weighing between 250 and 310 g per box. (Boxes weighing less than 250 g can lead to government fines, while those over 310 g lead to excessive costs.) If the process mean is 275 g, the standard deviation is 20 g and the process is normally distributed, find
   a. Percentage of boxes weighing less than 250 g.
   b. Percentage of boxes weighing more than 310 g.
   c. Revised mean to ensure that no more than 5% of the boxes will weigh less than 250 g.
   d. What is the new percentage of boxes that weigh more than 310 g?
   e. What would the revised high limit have to be to ensure that no more than 5% of boxes will weigh more than this new limit (use the revised mean)?

   Answers: 0.1057; 0.0401; 282.9; 0.0877; 315.8

9. Assume the mean and standard deviation of Problem 8 were derived from a sample of 10 (the *t* distribution must now be used). Find a through e.

   Answers: 0.1266; 0.0592; 286.66; 0.1432; 323.32

10. Find the answers to Problem 7 if the mean and standard deviation had been derived from a sample of 14 (use the *t*).

   Answers: 17.29; 17.29; 65.42

11. Find the new mean in Problem 10 so that no more than 10% of the tubes will fail before 1900 hr.

   Answer: 2035 hr

12. Engineering specifications on the inside diameter of a certain bearing are 2.000 ± 0.005 in. The mean and standard deviation (estimated from similar products from similar processes) are assumed to be 2.0019 and 0.0027, respectively. Find the expected percentage scrap and percentage rework (assume that all undersized bearings can be reworked).

   Answers: 12.55%; 0.53%

13. What percentage of tubes in Problem 11 will last longer than 2100 hr?

   Answer: 26.63%

14. Find the new mean in Problem 12 so that no more than 1% scrap is produced. How are these types of means normally revised? What is the new percentage rework?

   Answers: 1.9987 in.; 8.53%

15. What must the revised mean be in Problem 12 to ensure that no more than 5% rework will be produced, on the average? What is the new percentage scrap?

    Answers: 1.99944 in.; 1.97%

16. Find the revised upper and lower specification limits for Problem 12 for a maximum of 1% scrap and 5% rework.

    Answers: 2.008 in.; 1.997 in.

17. In Problem 12, assume a cost of $1 per part reworked and $10 per part scrapped. Compare the incremental costs for Problems 12, 14, 15, and 16. Which method is the least costly?

    Answers: $1.26; $0.953; $0.2480; $0.1500

18. What must be done in Problem 12 to ensure that no more than 1% of product will be scrapped and no more than 5% will be reworked (94% between the limits)?

3 PROBABILITY THEOREMS

The purpose of this chapter is to present the fundamental theorems of probability. These theorems are useful in the calculation of many reliability values.

Objectives

1. Define probability and the symbols used in probability theory.
2. Define the seven fundamental probability laws and know how to use them in solving problems in probability.
3. Understand the three counting rules and know how to use them to calculate amounts, sets, and subsets of objects.

3.1 Definitions and Symbols

Probability can be defined as the chance that something will happen. Other terms (along with chance) that can be considered synonyms to probability are likelihood and tendency. Likelihood can be substituted for the word chance in the above sentence, but tendency is used in a slightly different context. "The tendency to be the same" or "the tendency to be different" are both probability measures.

Probability is measured in terms of a ratio; that is, one value divided into part of that value. Therefore, probability values range between 0.00 and 1.00. The total of all possible chances in a situation is equal to 1.00. Probability is also very frequently measured in percent, but percent is just the ratio multiplied by 100. (A ratio of 0.015 is equal to 1.5%.) The basic rules of probability apply to all probability and statistical distributions, even the normal. However, the basic probability laws apply more obviously to the discrete situation than to the continuous, and the basic discrete probability distributions are more easily derived from the basic probability rules.

There are only three types of symbols needed in probability theory. $P(A)$ or P_A are the symbols used for the probability of an event. These two symbols [$P(A)$ and P_A] mean the same thing, and can be interchanged with each other. Of course, numbers can be substituted for A [$P(1)$ or P_1] or other letters can be used (P_B, P_C, etc.). The lower case s is used to denote a success, a desirable event. In reliability, it is failures that are measured and evaluated and so s, in this case, stands for the number of failures. The lower case n is used to denote the total number of possibilities. This is identical in meaning to the sample size n used in the normal curve. The probability equation, then, is

$$P(A) = P_A = s/n$$

where

$$P_A = P(A) = \text{the probability of an event}$$

$$s = \text{the number of successes (or failures,} \\ \text{depending on what is being evaluated)}$$

$$n = \text{the total possible number of cases}$$

Example 3.1 In the flip of a coin there are two possible cases, a head or a tail. Thus, $n = 2$. There is one possible way of getting a head. Therefore, $s = 1$. The probability of getting a head, then, on one flip of a coin is $1/2$:

$$P(H) = s/n = 1/2$$

3.2 Probability Theorems

There are seven fundamental probability theorems. All are useful in reliability at one time or another; some more so than others.

Rule 1 The probability of an event lies between 0 and 1. Zero probability (0.00) is the certainty that event A will not occur. One (1.00) is the certainly that event A will occur. The formula is

$$P_A = P(A) = 0.00 \text{ to } 1.00$$

Rule 2 *The sum of the probabilities* of a situation is equal to 1.00. The equation is

$$P_1 + P_2 + P_3 + \cdots + P_n = 1.00$$

Example 3.2 In the flip of a coin, the probability of a head is 1/2 and the probability of a tail is 1/2. The probability of a head or a tail (all possible cases) is equal to 1.00 (1/2 + 1/2) (neglecting the infinitesimal probability of the coin landing on its edge).

$$P(H \text{ or } T) = 1/2 + 1/2 = 1.00$$

Example 3.3 In the roll of a die, there are six faces and therefore six possible outcomes ($n = 6$). Each of the numbers (1 to 6) occur only once, therefore, $s = 1$. The probability of any number from 1 to 6 on one roll of a die is 1.00:

$P(1, \text{ or } 2, \text{ or } 3, \text{ or } 4, \text{ or } 5, \text{ or } 6)$

$$= 1/6 + 1/6 + 1/6 + 1/6 + 1/6 + 1/6 = 1.00$$

Note: Rule 2 is used constantly in solving real life probability problems. It is critical in solving these problems to first determine what constitutes a set of all possible cases. For instance, in a sample size of three where the number of defectives are being counted, the set of all possible cases would be 0, 1, 2, and 3 defectives in the sample. The sum of their probabilities must equal 1.00.

Rule 3 The *complementary law* states that if P_A is the probability that an event will occur, then $1 - P_A$ is the probability that the event will not occur. The probability of it occurring plus the probability of it not occurring always equals 1.00.

Example 3.4 In the roll of a die, the probability that a 1 will occur is 1/6. The probability that it will not occur is $1 - 1/6 = 5/6 = 83.3\%$.

Rule 4 In the *additive law* of probability, the probability of either A or B is the sum of the probability of A and the probability of B. In this law the two events must be *mutually exclusive*, that is, the occurrence of one event makes the other impossible (cannot have both a head and a tail on one toss of a coin). The equation is

$$P(A \text{ or } B) = P_A + P_B$$

The "or" is very distinctive in this case and always means plus, or add (except in the case of the combination law where it can also mean minus). Where the additive law is applicable, the "or" will always occur in the statement of the problem (or be implied in the logic).

Example 3.5 The probability of a head or a tail in one toss of a coin is 1.00 (neglecting the infinitesimal probability that the coin will land on its edge):

$$P(H \text{ or } T) = P(H) + P(T) = 1/2 + 1/2 = 1.00$$

Example 3.6 What is the probability of a 1 or a 6 occurring on one roll of a die?

Solution:

$$P(1 \text{ or } 6) = P(1) + P(6) = 1/6 + 1/6 = 2/6 = 0.333$$

Example 3.7 If the probability of obtaining one defective in a sample of 6 is 0.09 and the probability of obtaining two defectives in a sample of 6 is 0.04, what is the probability of obtaining one or two defectives in the sample?

Solution:

$$P(1 \text{ or } 2) = P_1 + P_2 = 0.09 + 0.04 = 0.13$$

Rule 5 The *multiplicative law* of probability states that the mutual probability of two independent events is equal to the product of the probabilities of each event. If two events are *independent*, the occurrence of one does not affect the probability of the other occurring (the occurrence of a head on the toss of one coin has no effect on the probability of a head or a tail occurring on a toss of a second coin). The equation is

$$P(A \text{ and } B) = P(A) \times P(B)$$

The "and" in this case always refers to multiplying, never to addition or subtraction. Where the multiplicative law is applicable, the "and" will always occur in a statement of the problem (or be implied in the logic).

Example 3.8 What is the probability that, in two tosses of a coin, the first toss will be a head and the second toss will be a tail?

Solution:

$$P(H \text{ and } T) = P(H) \times P(T) = 1/2 \times 1/2 = 1/4$$

Example 3.9 What is the probability that, in two tosses of a coin, there will be a head and a tail (in any order)? Note that in the problem statement, an "or" is also implied, besides an "and." Therefore, both the multiplicative and the additive laws apply. Two conditions will satisfy the solution of the problem: (1) a head on the first coin *and* a tail on the second *or* (2) a tail on the first *and* a head on the second.

Solution:

$$P[(H \text{ and } T) \text{ or } (T \text{ and } H)] = [P(H) \times P(T)] + [P(T) \times P(H)]$$

$$= (1/2 \times 1/2) + (1/2 \times 1/2) = 1/4 + 1/4 = 1/2$$

Note: Another way of analyzing this problem is to calculate the probabilities of all possible results. First note that only four combinations are possible: (1) A head on the first toss and a tail on the second; (2) a head on the first toss and a head on the second; (3) a tail on the first toss and a head on the second; and (4) a tail on the first toss and a tail on the second. Since these four combinations include all possible results, the sum of their probabilities equal 1.00. The probabilities of these four combinations are summarized as follows:

1. $P(H \text{ and } T) = P(H) \times P(T) = 1/2 \times 1/2 = 1/4$
2. $P(H \text{ and } H) = P(H) \times P(H) = 1/2 \times 1/2 = 1/4$
3. $P(T \text{ and } H) = P(T) \times P(H) = 1/2 \times 1/2 = 1/4$
4. $P(T \text{ and } T) = P(T) \times P(T) = 1/2 \times 1/2 = \underline{1/4}$

$$\text{Total} \quad 1.00$$

Now Example 3.9 can be answered by noting that two of the four combinations will solve the problem; the $P(H \text{ and } T)$ or the $P(T \text{ and } H)$. Since either one or the other will solve the problem, the solution is to add the two probabilities together ($1/4 + 1/4 = 1/2$). If, on the other hand, the question had been stated as in Example 3.8 (what is the probability that the *first* toss will be a head and the *second* will be a tail), only one of the above combinations would have answered the problem [$P(H \text{ and } T)$] and thus the answer would be $1/4$. It is imperative, in these types of problems, to analyze carefully and to apply carefully the applicable rule or rules. Note that an exhaustive analysis, such as was presented in this paragraph, is seldom feasible except when the number of possible outcomes is small.

Example 3.10 The probability that the first order of screws will be rejected is 0.12. The probability that the second order will be rejected is 0.30. What is the probability that they will both be rejected?

Solution:

$$P(A \text{ and } B) = P(A) \times P(B) = 0.12 \times 0.30 = 0.036$$

Rule 6 The *combination law* is used to find the probability of occurrence of either one or both of two events. The equation is

$$P(A \text{ or } B \text{ or both}) = P(A) + P(B) - [P(A) \times P(B)]$$

Note that both the additive and then the multiplicative laws are combined in this rule.

This formula can be generalized to three or more events. However, when three or more events are used, the term subtracted must include all events in the product. If it does not, the multiplicative law cannot be used in this subtracted term and the amount subtracted must be determined by logic alone.

Example 3.11 In a deck of 52 playing cards, what is the probability of getting a jack or a diamond on one draw? Note that the probability of getting a jack includes the probability of one diamond while the probability of a diamond includes the probability of one jack. Therefore, when the two probabilities are added (P_J and P_D), the probability of getting a jack of diamonds is included twice. One of them, therefore, must be subtracted. Since the probability of a jack of diamonds is 1/52, this figure could have been deduced by the above logic alone rather than calculated with the multiplicative law. In this case, the term subtracted would have been P_{JD} not ($P_J \times P_D$).

Solution:

$$P(J \text{ or } D) = P(J) + P(D) - (P_J \times P_D)$$

$$= 4/52 + 13/52 - (4/52 \times 13/52)$$

$$= 17/52 - 1/52 = 16/52 = 0.308$$

Note: An exhaustive analysis can also be made on this problem. Although this type of analysis is very inefficient in this case (there are so many possible outcomes), it will nevertheless be presented for illustration and instructional purposes. Note that the probability of drawing a jack or a diamond is equal to the probability of drawing a jack of diamonds or a jack of hearts or a jack of clubs or a jack of spades or a 2 of diamonds or a 3 of diamonds or a 4 of diamonds or a 5 of diamonds or a 6 of diamonds or a 7 of diamonds or an 8 of diamonds or a 9 of diamonds or a 10 of diamonds or a queen of diamonds or a king of diamonds or an ace of diamonds. (Note that the jack of diamonds is included only once in this list.) The probability then is the sum of the individual probabilities of each item on the list. But each item is just one card and the probability of drawing any one card from a deck of 52 cards is 1/52 — therefore, the sum of these 16 items is 16/52, or the $P(J \text{ or } D) = 16/52 = 0.308$.

The subtracted term in the combination law ($P_A \times P_B$) is not always obvious or easy to derive. In fact, the term ($P_A \times P_B$) can only be used when all items are equal in amount. For instance, there are four jacks, four queens, etc., and there are 13 of each of the four suits. If these variables were not equal (say only three jacks but four of each of the others), then the subtracted term must be determined by logic alone (as the P_{JD} was determined). This problem becomes especially difficult when the law is generalized to more than three events or occurrences.

Rule 7 The *conditional law* of probability applies to dependent events. Two events are dependent if the occurrence of one affects the probability of the second occurring — it does not necessarily affect the occurrence of the second event, just the probability that it will occur. The equation is

$$P(A \text{ and } B) = P(A) \times P(B|A)$$

$P(B|A)$ means the probability of B occurring given that A has already occurred (used only in the special case where the previous occurrence of A affects B's probability). $B|A$ is usually referred to as B given A.

Example 3.12 A tote box contains 60 shafts of which eight are defective. If a sample of two is removed, what is the probability that both will be defective (assume that neither shaft is replaced)? Note that after the first shaft is removed, the probabilities change, since there are only 59 shafts remaining. If the first one removed is a defective one, this again changes the probabilities since only seven defectives remain.

Solution:

$$P(A \text{ and } B) = 8/60 \times 7/59 = 0.016$$

Note: If each shaft had been measured and then replaced, the probability of getting a defective shaft on the second draw would have been independent of the first draw. The multiplication law would have applied in this case and the answer, then, would be

$$P(A \text{ and } B) = 8/60 \times 8/60 = 0.018$$

Independence usually leads to higher probabilities.

These seven laws can all be generalized to more than two occurrences. Care must be taken, however, in applying them to more than two occurrences. This is especially true of the combination law and the conditional law. These laws when applied to three or more events can be very complex and require much thought and analysis. The laws can also be combined in problems, and often are (see Examples 3.9 and 3.11). In fact, the combination law is just a special case of the combined multiplicative and additive laws.

3.3 Counting Rules

It is frequently important in reliability and probability to determine the number of sets and subsets of objects. There are three main methods to do this, each having application to particular types of sets.

Simple Multiplication

When it is desired to know the total number of possible sets, the rule of simple multiplication usually applies. If event A can happen in any of n_1 ways and event B can happen in any of n_2 ways, the total number of ways that both can occur is n_1 times n_2.

Example 3.13 There are five possible inspection routes in the morning and six possible in the afternoon. What is the total number of daily inspection routes possible?

Solution:

$$\text{Total} = n_1 n_2 = 5 \times 6 = 30$$

Example 3.14 An electronic system has two components. Component A has three different parallel circuits and component B has four. How many different ways can the current travel through the system?

Solution:

$$n_1 n_2 = 3 \times 4 = 12$$

If component C with five parallel circuits is added to the system, how many ways are there now?

Solution:

$$n_1 n_2 n_3 = 3 \times 4 \times 5 = 60$$

Permutations

If the possible sets of objects are to be ordered (arranged in specific ways), then these sets are called permutations. A permutation is defined as an ordered arrangement of n objects taken i at a time. Suppose that three letters of the alphabet (A, C, and T) are chosen, and it is desired to arrange them in definite order. In other words, the order of appearance of the letters is important (this is the case, for instance, when letters of the alphabet are arranged to form words). Permutations of these three letters, then, are ACT, ATC, TAC, TCA, CAT, and CTA. There are six possible permutations of these three letters (even though only two make recognizable words). The equation is

$$P_i^n = \frac{n!}{(n-i)!}$$

Note: The exclamation mark (!) means factorial: $n! = (n)(n-1)(n-2)(n-3) \cdots (1)$; and $0! = 1$.

Example 3.15 Find the permutation of five things taken two at a time.

Solution:

$$P_2^5 = \frac{5!}{(5-2)!} = \frac{5 \times 4 \times 3 \times 2 \times 1}{3 \times 2 \times 1} = 5 \times 4 = 20$$

Combinations

If the way the objects are ordered is unimportant, the set is called a combination. Using the same three letters of the alphabet used in the section on permutations (A, C, and T), there is only one combination. Since the same three letters are used, and the order of arrangement does not matter, then (as far as combinations go) ACT = TAC = CAT, etc. There are always more permutations than combinations in the same set of objects. (The letters A, C, T have six permutations, but only one combination, except for the null set of n things taken 0 at a time when the permutations = the combinations = 1.) The equation for a combination is

$$C_i^n = \frac{n!}{i!(n-i)!}$$

Example 3.16 Find the combination of five things taken two at a time.

Solution:

$$C_2^5 = \frac{5!}{2!(5-2)!} = \frac{5 \times 4 \times 3 \times 2 \times 1}{(2 \times 1)(3 \times 2 \times 1)} = \frac{5 \times 4}{2} = 10$$

Terms and Definitions

1. **Event.** A measurable condition.
2. **Success.** A desirable event $= s$.
3. **Failure.** An undesirable event $= f$.
4. **Finite.** Having a beginning and an end.
5. **Probability theorems.** Generally accepted laws that govern probability calculations.
6. **Additive law.** $P(A \text{ or } B) = P(A) + P(B)$.
7. **Multiplicative law.** $P(A \text{ and } B) = P(A) \times P(B)$.
8. **Dependent/independent.** One event does/does not affect the probability of the other occurring.
9. **Mutually exclusive.** Both events cannot happen at the same time.
10. **Conditional theorem.** $P(A) \times P(B|A)$.
11. **Permutation.** An ordered set of objects.
12. **Combination.** An unordered set of objects.

Practice Problems

1. A single ball is to be selected at random from a container containing a mixture of several different colored balls. The probability of selecting a red ball is 0.27, a green ball is 0.13, a yellow ball is 0.35, and a pink ball is 0.22. Although there are a few purple balls in the

container, their probability is unknown. What is the probability that the ball is either green or yellow? Green or purple? Green or red or pink?

Answers: 0.48; 0.16; 0.62

2. Four different types of shafts, all numbered for identification, are mixed in a tote box. There are eight straight shafts numbered 1 to 8, 12 tapered shafts numbered 1 to 12, 6 stepped shafts numbered 1 to 6, and 4 square shafts numbered 1 to 4. If one shaft is to be selected at random, what is the probability of selecting a square shaft or a shaft numbered four? A tapered shaft or a shaft numbered 7? A stepped shaft or a shaft numbered 5?

Answers: 0.2333; 0.4333; 0.2667

3. A basket contains 55 parts of which 12 are known to be defective. If a sample of two is drawn at random and not replaced, what is the probability that both are defective? What is the probability that a sample of three (that are not replaced) are all defective? What is the probability that, in a sample of two that are drawn at random and not replaced, one is defective and one is not? What is the probability that the first part is good and the second is defective in a sample of two drawn at random and not replaced?

Answers: 0.0444; 0.0084; 0.3475; 0.1737

4. A delivery truck can travel three different routes on Monday, four on Tuesday, five on Wednesday, six on Thursday, and seven on Friday. If a route is chosen at random each day, how many different weekly routes are possible?

Answer: 2520

5. A sample of four is selected from a lot of 40. How many permutations are possible? How many combinations are possible?

Answers: 2,193,360; 91,390

6. A sample of 10 is selected from a lot of 20 parts. How many permutations are possible? How many combinations are possible?

Answers: 6.70442×10^{11}; 184,756

4 | # THE
BINOMIAL
DISTRIBUTION

Sample Size
Pass/Fail Data

This chapter introduces the concept of discrete distributions in general and the binomial in particular. The binomial forms the theoretical basis for all discrete distributions.

Objectives

1. Define the binomial and understand its nature, including the concept of a discrete distribution; the symbols p, q, and n; and the relationship to the normal curve.
2. Understand the use of the binomial expansion in probability.
3. Understand the use of the single-term binomial formula and know how to use it to solve discrete probability problems.

4.1 The Nature of the Binomial

The binomial is a discrete distribution where the sum of its components is equal to 1.00. Discrete means that the individual values are distinctive and do not blend into each other. They are attributes that cannot be divided (unlike continuous variables that have an infinite number of divisions between any two measurements). They are represented by the counting numbers 1, 2, 3, etc. Discrete variables are counted, while continuous variables are measured. The number of failures in a test is a discrete value, since these are counted and represented by whole numbers. (You cannot have half a failure, or half a defective, for instance.) Each term in the binomial represents the probability of a particular number of successes (or failures, depending on what is being measured) and the sum of all the terms is equal to 1.00 (the sum of all possible cases = 1.00). (The first term is p_n; the second term is p_{n-1}; etc.).

The binomial is a dichotomous distribution, with its two parts each representing an either/or situation (good or bad, success or failure, pass or fail, etc.). The two parts to the binomial are represented by the symbols p and q, with p representing a success and q a nonsuccess. A success, of course, can be any desired (or expected) condition, even a failure. Since only two conditions are possible (*bi* meaning two) in the binomial, the sum of the probabilities of the two conditions must equal to 1.00. Thus, $p + q = 1.00$ and $q = 1 - p$. The binomial can approximate the normal curve, and vice versa, and, when it does, the mean of the binomial is np and the standard deviation is $\sqrt{npq}$:

$$\bar{x} = np$$

$$s = \sqrt{npq}$$

4.2 The Binomial Expansion

There are two ways, formulas, that can be used to calculate the binomial; the binomial expansion and the single-term formula. The equation for the binomial expansion is

$$(p + q)^n = p^n + np^{n-1}q + \frac{n(n-1)}{2} p^{n-2}q^2$$
$$+ \frac{n(n-1)(n-2)}{3!} p^{n-3}q^3 + \cdots + q^n$$

where

$$p = \text{probability of an event}$$

$$q = \text{probability of a nonevent } (1 - p)$$

$$n = \text{number of trials or sample size}$$

In this expansion, each term represents the probability of a particular number of successes (or failures, depending on what is desired or expected — what is being measured). Thus, the first term p^n is the probability of n successes; the second term is the probability of $n - 1$ successes (p^{n-1}), etc. There are always $n + 1$ terms in the binomial representing the probability of n, $n - 1$, $n - 2$, ..., 0 successes, respectively. Note that the final term is the probability of zero successes (p^0 is, of course, assumed in this term).

4.3 The Single-Term Formula

Each of the terms of the preceding expansion can be represented by one formula:

$$p(i) = C_i^n p^i q^{n-i}$$

where

$p(i)$ = probability of i successes

C_i^n = coefficient = combination of n things taken i at a time

p = probability of an event

q = probability of a nonevent = $1 - p$

i = number of successes in n trials

n = number of trials or sample size

This formula is extremely useful, since it allows the calculation of an individual probability without having to first expand the binomial (a tedious process for large sample sizes).

Example 4.1 A random sample of 10 is selected from a steady stream of product from a punch press, which past experience has shown to produce 10% defective parts. Find the probability of (1) one bad part, (2) two bad parts, (3) one or less bad parts, and (4) three or more bad parts in the sample.

Solution:

(1) One bad part:

$$P(i) = C_i^n p^i q^{n-i}$$

$$P(1) = C_1^{10}(0.10)^1(0.90)^{10-1}$$

$$= \frac{10!}{1!\,(10 - 1)!}\,(0.10)^1(0.90)^9$$

$$= 10\,(0.10)\,(0.3874)$$

$$= 0.3874 \text{ or } 38.74\%$$

(2) Two bad parts:

$$P(2) = \frac{10!}{2!\,(10 - 2)!}\,(0.10)^2(0.90)^8 = 0.1937 \text{ or } 19.37\%$$

(3) One or less bad parts:

$$P(1 \text{ or less}) = P(0) + P(1) \qquad \text{[additive law]}$$

$$P(0) = \frac{10!}{0!(10 - 0)!}\,(0.10)^0(0.90)^{10} = 0.3487 \text{ or } 34.87\%$$

$$P(0) + P(1) = 0.3487 + 0.3874 = 0.7361 \text{ or } 73.61\%$$

(4) Three or more:

$$P(3 \text{ or more}) = 1 - P(2 \text{ or less}) \qquad \text{[complementary law]}$$

In the left margin (handwritten): *This is sampling w/ replacement. Sampling w/o replacement, use hypergeometric distribution.*

$$= 1 - (P_0 + P_1 + P_2) = 1 - (0.3487 + 0.3874 + 0.1937)$$

$$= 1 - 0.9298 = 0.0702 \text{ or } 7.02\%$$

Note how useful the probability laws are in solving binomial problems. They are equally as useful with the Poisson distribution (Chapter 5).

Derivation of the Binomial

The binomial is derived from the *product* rule of probability (a special case of the *conditional* rule, where the values are independent). The derivation will be explained in conjunction with the following example.

A lot of 20 is known to have five defective parts. What is the probability that a sample of four, taken with replacement (each part is chosen at random, measured, and then replaced before the next part is chosen), will have exactly one replacement?

For the binomial, this problem would have an identical solution if the lot size were unknown and the sample were taken from a process known to produce 25% defective parts, on the average (in other words, for the binomial, the two problems are identical).

Solution:

1. Any sample, of four, that is a success (satisfies the problem statement) will have one bad part and three good parts.
2. Since their probabilities are independent, the probability of getting a bad part (5/20) will always be the same, no matter where it appears in the sample. The probability of getting a good part (15/20) will also be the same no matter where it appears in the sample.
3. There are four samples, out of all possible samples, that will satisfy these conditions:
   a. *B, G, G, G*
   b. *G, B, G, G*
   c. *G, G, B, G*
   d. *G, G, G, B*
4. The probability that sample "a" will occur is: $5/20 \times 15/20 \times 15/20 \times 15/20 = 0.1055$.
5. Since any mixture of the same four fractions, when multiplied together, will always have the same result, the probability of any one of the above four samples will always be the same, 0.1055.
6. Since any one of the preceding four samples will satisfy the problem conditions (is a success), and no other, the probability of one defective in a sample of four is: $P(1) = P(a) + P(b) + P(c) + P(d) = 0.1055 + 0.1055 + 0.1055 + 0.1055 = 4 \times 0.1055 = 0.422$.

The number of ways that a desired condition can occur is the combination of four things taken one at a time (since the order of occurrence is unimportant).

An equation for the preceding problem, then, is

$$P(1) = C_1^4 p^1 q^{4-1}$$

Finally, this can be generalized into the single-term formula explained in the previous section:

$$P(i) = C_i^n p^i q^{n-i}$$

A Comparison of the Single-Term Formula to the Binomial Expansion

If $n = 5$, the binomial expands as follows:

$$(p + q)^5 = p^5 + np^4q + \frac{n(n-1)}{2!} p^3q^2 + \frac{n(n-1)(n-2)}{3!} p^2q^3$$

$$+ \frac{n(n-1)(n-2)(n-3)}{4!} p^1q^4 + \frac{n(n-1)(n-2)(n-3)(n-4)}{5!} p^0q^5$$

In comparison, the single-term formula becomes $_5C_i p^i q^{5-i}$ (where i is the expected or desired number of defective parts). This formula can be used in place of each of the terms in the above binomial expansion.

Example 4.2 Suppose that an assembly line normally produces 15% scrap parts. What is the probability that, in a sample of 5, there will be five defective parts? 4? 3? 2? 1? 0?

Solution (using the binomial expansion):

$$p = 0.15 \text{ and } q = 1 - p = 1 - 0.15 = 0.85$$

Substituting into the binomial expansion:

1. $p(5) = p^5 = (0.15)^5$ $\qquad\qquad\qquad$ $= 0.000076$

2. $p(4) = np^4q = 5(0.15)^4(0.85)$ $\qquad\quad$ $= 0.002152$

3. $p(3) = \dfrac{n(n-1)}{2!} p^3q^2$

    $= \dfrac{5 \times 4}{2} (0.15)^3(0.85)^2$ $\qquad\qquad$ $= 0.024384$

4. $p(2) = \dfrac{n(n-1)(n-2)}{3!} p^2q^3$

    $= \dfrac{5 \times 4 \times 3}{3 \times 2} (0.15)^2 (0.85)^3$ $\qquad\quad$ $= 0.138178$

5. $p(1) = \dfrac{n(n-1)(n-2)(n-3)}{4!} pq^4$

$= \dfrac{5 \times 4 \times 3 \times 2}{4 \times 3 \times 2}(0.15)(0.85)^4 \qquad = 0.391505$

6. $p(0) = \dfrac{n(n-1)(n-2)(n-3)(n-4)}{5!} q^5$

$= \dfrac{5 \times 4 \times 3 \times 2 \times 1}{5 \times 4 \times 3 \times 2 \times 1}(0.85)^5 \qquad = \underline{0.443705}$

Total   1.0000

Note: There are six terms, one more than the sample size, that total to 1.00000. The first term is the probability that the entire sample (in this case 5) is defective, the second term is the probability that all but one part are defective, and so forth. Note that the last term is always the probability that there are no (zero) defectives in the sample.

This example can also be solved using the single-term formula. The answers are given below for the sake of comparison. Note how the coefficient in each case (the C_i^n) reduces to match exactly the respective coefficients in the binomial expansion.

1. $p(5) = C_5^5 p^5 q^{5-5} = \dfrac{5!}{5!(5-5)!}(0.15)^5(0.85)^0$

$= \dfrac{5!}{5!0!}(0.15)^5 \qquad\qquad = 0.000076$

2. $p(4) = C_4^5 p^4 q^{5-4} = \dfrac{5!}{4!(5-4)!}(0.15)^4(0.85) \qquad = 0.002152$

3. $p(3) = C_3^5 p^3 q^{5-3} = \dfrac{5!}{3!(5-3)!}(0.15)^3(0.85)^2 \qquad = 0.024384$

4. $p(2) = C_2^5 p^2 q^{5-2} = \dfrac{5!}{2!(5-2)!}(0.15)^2(0.85)^3 \qquad = 0.138178$

5. $p(1) = C_1^5 p^1 q^{5-1} = \dfrac{5!}{1!(5-1)!}(0.15)(0.85)^4 \qquad = 0.391505$

6. $p(0) = C_0^5 p^0 q^{5-0} = \dfrac{5!}{0!(5-0)!}(0.85)^5 \qquad = \underline{0.443705}$

Total   1.00000

Terms and Definitions

1. **Binomial.** A discrete probability distribution giving the probability of obtaining a specified number of successes in a finite set of independent trials in which the probability of a success remains the same from trial to trial.
2. **Discrete.** Countable data — separate, distinct data, usually represented by whole numbers only; discontinuous data.
3. **Continuous.** Measurable data — uninterrupted data where an infinite number of divisions (theoretically) can be made between any two pieces of data.
4. **Dichotomous.** Data divided into two parts.

Practice Problems*

1. Using the binomial distribution, find the probability of obtaining two or more defectives when sampling five parts from a batch known to be 6% defective.
   Answer: 3.2%

2. Using the binomial distribution, find the probability of obtaining two or less defectives in a sample of nine when the lot is 15% defective.
   Answer: 86%

3. Find the probability of less than two defectives in Problem 1.
   Answer: 96.8%

4. Find the probability of exactly three in Problem 1.
   Answer: 0.19%

5. Find the probability of exactly three defectives in Problem 2.
   Answer: 10.69%

6. Find the probability of more than three defectives in Problem 2.
   Answer: 3.4%

7. Find the probability of three or more defectives in Problem 2.
   Answer: 14%

*These problems are all binomially distributed.

CHAPTER 5 THE POISSON DISTRIBUTION

This chapter introduces the Poisson distribution. This distribution is one of the most used probability methods in all reliability, owing to its almost universal application and ease of use.

Objectives

1. Define the Poisson and understand its nature and symbols.
2. Understand the Poisson formula and how to use it in determining probabilities.
3. Understand the use of the Poisson tables.
4. Understand the inverse use of the Poisson tables and how to use them to determine sample size and desired process average from a given probability.

5.1 The Nature of the Poisson

When the fraction defective (p) is small, the binomial distribution takes the form of the Poisson distribution. The Poisson distribution is most useful in determining probabilities for observations (defects or failures) per unit of time, area, or amount. Since reliability is concerned with small number of failures per unit of time, the Poisson is most appropriate. Generally, the Poisson distribution can be used to approximate the binomial distribution when p is small and n is quite large (or, in reliability, when failure rate is small and test times are relatively large).

The Poisson distribution, like the binomial distribution, can be used to approximate the normal distribution, when the mean ($\bar{x}$) is equal to np (or, in reliability, λT) and the standard deviation (s) is equal to $\sqrt{np}$ ($\sqrt{\lambda T}$ in reliability):

$$\bar{x} = np = \lambda T$$

$$s = \sqrt{np} = \sqrt{\lambda T}$$

As np gets larger, the Poisson distribution becomes more symmetrical — approaches the normal curve — and the normal distribution must then be used instead of the Poisson distribution. In reliability, however, this very rarely happens except in the case of break-in times for a new system (or mission) or in the case where a long-term system has reached the wear-out stage.

5.2 The Poisson Formula

The Poisson distribution is similar to the binomial distribution in that it consists of a number of terms, each term representing the probability of a particular number of occurrences, with the sum of all the terms equal to 1.00. The first term represents the probability of zero defects, the second term respresents the probability of one defect, etc. Unlike the binomial distribution, there is no end, theoretically, to the number of terms. However, the probabilities quickly become so small as to be infinitesimal and so, practically speaking, can be ignored. Thus, the Poisson tables give the probabilities of only a limited number of the Poisson terms (to three or four places).

The equation for the Poisson distribution is

$$P(c) - \frac{(np)^c}{c!} e^{-np}$$

where

$P(c)$ = probability of c defects or failures [identical to the $P(i)$ of the binomial distribution]

c = number of defects or failures that can be tolerated (usually determined by engineering or management)

p = fraction defective (as in the binomial distribution)

n = the sample size

np = average number of defects or failures (determined by actual sampling)

e = the base of the Naperian logorithms = 2.71828 . . .

A word of caution about the nature of c as compared to np. Since these two terms have similar meanings, it is easy to confuse them. However, remember that c is an individual value set by engineering, while np is the average of the distribution and is usually determined by actual sampling (occasionally, np is assumed). Even though np and c do have similar meanings, they can never be exchanged in the formulas.

Example 5.1 Ten radio sets were inspected and found to have 1, 0, 2, 3, 1, 2, 1, 0, 0, 0 defects, respectively. Calculate the average number of defects per set (np) and the probability of occurrence of 0, 1, 2, 3, 4, or 5 defects in any one set chosen at random. The average number of defects per set (np) is 1 (10/10).

Solution:

Number of defects (c)	Calculations	p
0	$\dfrac{1^0}{0!} 2.71828^{-1} = \dfrac{1}{1} \times \dfrac{1}{2.71828}$	= 0.368
1	$\dfrac{1^1}{1!} 2.71828^{-1} = \dfrac{1}{1} \times \dfrac{1}{2.71828}$	= 0.368
2	$\dfrac{1^2}{2!} 2.71828^{-1} = \dfrac{1}{2} \times \dfrac{1}{2.71828}$	= 0.184
3	$\dfrac{1^3}{3!} 2.71828^{-1} = \dfrac{1}{6} \times \dfrac{1}{2.71828}$	= 0.061
4	$\dfrac{1^4}{4!} 2.71828^{-1} = \dfrac{1}{24} \times \dfrac{1}{2.71828}$	= 0.015
5	$\dfrac{1^5}{5!} 2.71828^{-1} = \dfrac{1}{120} \times \dfrac{1}{2.71828}$	= 0.003
	TOTAL	= 0.999

An analysis of the Example 5.1 reveals the following:

1. The total of the six probabilities do not quite equal 1.00 because the distribution never ends (the number of defects can be infinite, theoretically). However, it would serve no useful purpose to continue the calculations because the remaining probabilities are effectively equal to zero (the probability of six defects, for instance, is 0.0005).
2. The "additive" law of probability is quite useful. For instance, the $P(1 \text{ or less})$ is equal to $P(0) + P(1) = 0.368 + 0.368 = 0.736$.
3. The "complementary" law of probability can also be useful. For instance, suppose it were desirable to know the probability of two or more defects. One way to calculate this would be to use the additive law and add all the probabilities above one. But this could be tedious and even, theoretically, impossible. Therefore, $P(2 \text{ or more}) = 1 - P(1 \text{ or less}) = 1 - 0.736 = 0.264$.

Note: In this example, the Poisson is applied to defects per unit. It is equally applicable to defective units in a sample. The sample can be drawn from a supplier, or from an infinite universe, such as product delivered continuously off the end of the production line. In the finite situation, the fundamental distribution is the hypergeometric. In the infinite situation, the binomial is the fundamental distribution. The Poisson can frequently be used to effectively approximate either of these distributions.

5.3 The Poisson Tables

The calculation of the Poisson formula, though possible, is tedious and time-consuming, especially when many terms are involved [like $P(5$ or less), for instance]. The Poisson tables (Table A2) have been prepared to overcome this problem. The tables are very easy to use, which is one reason the Poisson distribution has found wide applications in many fields.

To use the tables, two values are needed: np and c. np is found in the left-hand column and c is across the top. The body of the table is the probability, and it is cumulative — it is an "or less than" value. The "additive" and "complementary" laws are especially useful in using these tables. A list of rules has been developed, using these laws, to assist the user (using $c = 3$ and $np = 1$ for illustration).

1. $P(3$ or less) − use the table direct. Find np in the left-hand column and move to the right to find the probability value under the proper c value. For example, if $np = 1.00$ and $c = 3$ or less, $P(3$ or less) $= 0.981$ or 98.1%.
2. $P(3) = P(3$ or less) − $P(2$ or less) $= 0.981 − 0.920 = 0.061$.
3. $P($more than 3) $= 1 − P(3$ or less) $= 1 − 0.981 = 0.019$.
4. $P(3$ or more) $= 1 − P(2$ or less) $= 1 − 0.920 = 0.080$.
5. $P($less than 3) $= P(2$ or less) $= 0.920$.

Example 5.2 A hospital had 500 cases of a certain disease during the year, of which five patients died. The national average for this type of disease was six out of 1000. What is the probability of having four or five deaths in 500 cases?

$p = 6/1000 = 0.006$ occurrences per case

$np = 500(0.006) = 3.0$ deaths per 500 cases (on the average)

Solution:

$P(4$ or 5) $= P(4) + P(5) = P(5$ or less) − $P(3$ or less)

$$= 0.916 − 0.647 = 0.269 \text{ or about } 27\%$$

Example 5.3 A new manager of an airport found that 18 airplanes could be safely landed per hour by the airport. Last year an average of 15 planes per hour had been landed. What is the probability that, in any 1-hr period, there will be more than 18 planes trying to land?

Solution:

$$p = \text{unknown}$$

$$n = \text{unknown}$$

$$np = 15 \text{ planes per hour (average)}$$

$$c = \text{more than } 18$$

$$P(\text{more than } 18) = 1 - P(18 \text{ or less}) = 1 - 0.819 = 0.181 \text{ or } 18.1\%$$

A new computer will now allow 25 planes to be safely landed each hour. What is the probability that more than 25 will be trying to land in any 1 hr?

$$P(\text{more than } 25) = 1 - P(25 \text{ or less}) = 1 - 0.994 = 0.006 \text{ or } 0.6\%$$

Note: Cost and safety factors would also be involved in the actual decision.

5.4 The Inverse Use of the Poisson Distribution

There are times when management would like to know what the required sample size should be for a given probability (or what the process average must be). To get these answers the formulas and rules already given must be worked backward. This is called the "inverse use of the Poisson," and the steps are:

1. Find the given probability in the body of the table under the appropriate c value:
   a. If $P(c \text{ or less})$, use the probability value and c value as given — go directly to the body of the table for P under c.
   b. If $P(c \text{ or more})$, go to the body of table for $1 - P$ under $c - 1$.
   c. If $P(\text{more than } c)$, go to the body of table for $1 - P$ under c.
2. Find np in the left-hand column (interpolate if necessary).
3. Calculate $n = np/p$ or $p = np/n$.

Example 5.4 What must the process average be for a sample of 10 if $P(3 \text{ or less}) = 0.891$?

Solution:

Find 0.891 in the body of Table A2 under $c = 3$. np for this value is 1.8. Therefore

$$p = np/n = 1.8/10 = 0.18 \text{ or } 18\%$$

Example 5.5 Find the sample size needed if the fraction defective is 0.03 and $P(2 \text{ or more}) = 0.236$.

Solution:

$$P(\text{2 or more}) = 1 - P(\text{1 or less}) = 0.236$$

$$P(\text{1 or less}) = 1 - 0.236 = 0.764$$

Find 0.764 in the body of Table A2 under $c = 1$ and interpolate to find $np = 0.922$:

$$n = np/p = 0.922/0.03 = 30.7 \text{ or } 31 \text{ units}$$

Example 5.6 Find the sample size necessary if the process average is 0.03 and $P(\text{more than 2}) = 0.236$.

Solution:

$$P(\text{more than 2}) = 1 - P(\text{2 or less}) = 0.236$$

$$P(\text{2 or less}) = 1 - 0.236 = 0.764$$

Find 0.764 in the body of Table A2 under $c = 2$ and interpolate to find $np = 1.6731$:

$$n = np/p = 1.6731/0.03 = 55.77 \text{ or } 56 \text{ units}$$

Terms and Definitions

1. **Poisson distribution.** A discrete probability distribution that is a limiting form of the binomial distribution for small values of the probability of success and for large numbers of trials.
2. **Inverse use of the Poisson distribution.** Working backward from a given table value to find an unknown sample size or fraction defective.

Practice Problems*

1. If the probability is 0.08 that a single article is defective, what is the probability that a sample of 20 will contain two or less defectives? Two or more?
   Answers: 0.783; 0.475

2. A steady flow of product has a fraction defective of 0.09. If 67 are sampled, what is the probability of three defectives? Three or less? Zero (0) defectives? More than three?
   Answers: 0.089; 0.151; 0.002; 0.849

3. $P(\text{3 or less defectives}) = 0.896$. What is the fraction defective (p) for a sample of 10?
   Answer: 0.177

*These problems are all Poisson distributed.

4. P(3 or more defectives) = 0.336. What is the sample size if the fraction defective is 0.10?

   Answer: 20

5. P(0 defectives) = 91% when the fraction defective is 0.01. What is the sample size?

   Answer: 9

6. A check-out station can serve five customers every 15 min. Last year, during the four peak hours of the day (5–9 pm), there were 30 customers per hour, on the average. What is the probability that a customer will have to wait more than 15 min? How many of 30,000 customers last year (during peak hours) had to wait longer than 15 min to be served? A computerized robot can double the number of customers served. How many of the 30,000 customers last year would have had to wait longer than 15 min if the robot had been operating during that year?

   Answers: 0.758; 22,740; 4125

7. A certain process normally runs 3% scrap. Out of a sample of 50, what is the probability that more than four will be scrap? Less than four? Only four? Four or less? Four or more? Out of 1000 samples of 50, how many samples will have more than three defects?

   Answers: 0.019; 0.934; 0.047; 0.981; 0.066; 66

8. Management wishes to have a 95% probability that there will be no more than two defectives per sample. (a) What should the sample size be if the process normally runs 4% scrap? (b) What should the process average be (actually the maximum process average) for a sample size of 10?

   Answers: 21; 0.082

9. Management wants no more than a 3% probability that the number of failures will be greater than one on any one sample. (a) What should the sample size be for a process that normally runs 1% scrap? (b) What should the process average be for a sample size of 20?

   Answers: 27; 0.0134

CHAPTER 6 | COMPONENT RELIABILITY

This chapter presents methods of analyzing failures and predicting reliability for single components and single systems. The exponential, binomial, normal, and Weibull distributions are used to calculate reliability (for a stated mission time), mission time (for a stated reliability), and a new mean (for a stated reliability and mission time).

Objectives

1. Understand the difference between quality control analysis of defectives and reliability analysis of failures.
2. Know the four main probability distributions that apply to reliability, and the three questions that they are used to solve.
3. Use the exponential to solve the three reliability questions.
4. Use the binomial to solve the three reliability questions.
5. Use the normal to solve the three reliability questions.
6. Use the Weibull to solve the three reliability questions and understand how to use the Weibull graph to solve reliability problems.
7. Compare the four distributions as to results and understand the interrelationships of mission time, mean life, and failure rate to reliability. Also know how to do a "goodness-of-fit" test to determine the correct distribution.
8. Understand the concepts and calculations of conditional reliability.

6.1 Quality Control versus Failure Analysis

In Chapter 5, the Poisson distribution was presented. However, the discussion in that chapter was limited to procedures and equations for analyzing defective parts. This is essentially a quality control type analysis and is used to predict the probability of occurrences of defects in a continuum of volume or area, to control production, and to design inspection sampling plans. The Poisson distribution can also be used to analyze and predict failures in a continuum of time (and thus forms the basis of reliability prediction), to design reliability sampling plans (Chapters 10 and 11), and to provide information to engineering for design of systems.

6.2 Probability Distributions for Reliability

There are four main probability distributions that can be used to determine the reliability of a component or a single system (a system that is treated as if it were a single component). They are:

1. Exponential.
2. Binomial.
3. Normal (Guassian).
4. Weibull.

The exponential and binomial distributions can be used with both time-terminated and failure-terminated data, but only failure-terminated data can be used with the normal and Weibull distributions, because time-terminated tests (where not all units fail) only produce one parameter, the mean life (or failure rate). The standard deviation and the Weibull parameters (B and G) require failure-terminated tests (where all units are tested to failure) for their calculations.

In the past, the exponential has been the most used distribution in all aspects of reliability, perhaps because it is also the easiest to use and because most complex systems exhibit exponential qualities (the mixture of break-in, constant, and wear-out failures, all present at the same time, causes the average failure rate to approximate a constant failure rate). However, the Weibull distribution has become more popular recently owing to its ability to approximate a wide range of different distribution types with one series of formulas (a relatively simple graph can also be used instead of the complex formulas).

The exponential and binomial are limited in that they can be used for the constant failure rate case only (see Fig. 1.1). If the failure rate is decreasing (break-in or infant mortality stage) or increasing (wear-out stage), the normal or Weibull distribution must be used. However, the normal and Weibull distributions theoretically can be used at any of the three stages, if the data fit the stage, of course. The Weibull distribution is especially useful, since it can be made to fit any of the three stages with most data distributions.

The binomial distribution has another problem that the exponential does not. Time units must be used that guarantee the failure rate will be less than 1.0. This is done by converting the data to a finer

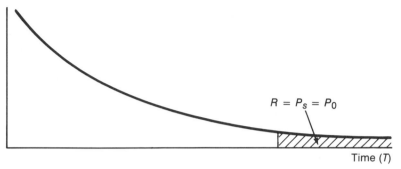

$$R = P_s = P_0$$

Time (T)

Fig. 6.1 Reliability for the exponential distribution.

time discrimination (change hours to minutes, for instance), but this is time consuming and adds another dimension for error. Actually, the exponential does an excellent job of approximating the binomial when the mission time (T) is large and the failure rate is relatively small (almost all reliability calculations are of this nature). Therefore, the exponential can be used, in most cases, in the place of the binomial. However, the binomial is being used somewhat in reliability and so will be covered in this chapter.

Some reliability texts, in the past, have promoted the use of the Poisson distribution for determining the probability that the system will survive one or more failures (where a few failures are allowed). The Poisson tables have been used at $c = 1, 2$, or more failures. This is theoretically incorrect. If a few failures can be allowed, and management (or the customer) wishes to know the probability that the system will last for the mission time, T hr, with no more than a predetermined number of failures (f), the mean life must be recalculated for that number of failures. This new mean life, then, is used exactly as if it were the mean life of a test where no failures were allowed. The formulas and procedures are identical. If the Poisson tables are used to determine the reliability in this case, only the $c = 0$ column must be used. In fact, the $c = 0$ column is the only one that should ever be used with reliability problems, no matter what procedures have been used to determine the mean life. Only the statement of the problem, and the problem solution, are changed to reflect the allowed failures (and the test for mean life is changed to last until the number of allowed failures has occurred). For instance, a reliability statement might say: "The product has a 90% probability of lasting for 100 hr with no more than two failures." Since reliability is the probability of zero failures for the mission time, it is theoretically incorrect to refer to this probability (of two or less failures in 100 hr) as reliability.

In Chapters 4 and 5 the distribution parameters were defined by the sample size n and the fraction defective p (or np). In most statistical analyses (quality control, for instance), sample size usually refers to a group of parts chosen at random from a larger lot (or from an on-going production process) and measured for a particular characteristic. Defectives are then determined by comparison to specification limits. In reliability, sample size and fraction defective have slightly different connotations. A reliability sample size refers instead to a sample of time units (hours, minutes, weeks, etc.), while a reliability fraction defective is the failure rate, λ. The important reliability time sample is the mission time, T, which is equivalent to the n of Chapters 4 and 5. The failure rate, λ, is equivalent to the p. Therefore, λT is a special kind of np and can be used in place of np in the formulas. λT is always an np, but np is not always a λT. Since m is the reciprocal of failure rate, $1/m$ can be substituted for λ in any of the formulas. In reliability, m (or MTBF) is often a more meaningful control value than failure rate and is easier to use in some of the formulas (it is also more intuitively obvious for most people).

In the remainder of this chapter, the four different distributions will be used, separately, to solve the following three reliability questions:

1. What is the reliability given a mission time (and failure rate or mean life data)?
2. What is the mission time given a desired reliability (and failure rate data)?
3. What must the mean life be (or what is the new mean) for a desired reliability and stated mission time? This question does not require a test and, so, can be used to determine what the mean life ought to be, and what the production process must be able to accomplish, even before the design is started.

6.3 The Exponential Failure Law

One of the most important values in reliability is the probability of survival, which is just the probability of zero failures. Obviously, given enough time, all things must fail. Therefore, the probability of survival (P_s or R), or its equivalent, the probability of zero failures (P_0), must have an associated time period. In this book, if a time period is not specifically stated, it is assumed (usually 100 hr). The time period can be ignored for purposes of illustration in this book, but not in real life.

The probability of zero failures is just the first term of the Poisson expansion. Because $c = 0$ in this term, it can be simplified to e^{-np} [$(np^0) = 1$ and $0! = 1$ so that $(np^0)/0!$ reduces to 1 and thus can be ignored]. In reliability, however, the sample size n is a time sample T, while the fraction defective p is changed to the failure rate λ. The equation then becomes

$$P_0 = P_s = R = e^{-\lambda T}$$

This is the equation of another distribution, called the exponential distribution, and, in this form, is called the exponential failure law. Since λ is equal to the reciprocal of the mean life ($\lambda = 1/m$), an alternate from of the above equation is

$$R = e^{-(1/m)T} \quad \text{or} \quad R = e^{-T/\overline{m}}$$

The T in the preceding formulas is a probability time sample and is usually determined by engineering requirements or management judgment. It has no relation to test period, average test time, or any other test time, although it is occasionally equal to the test period.

The exponential failure law can be found from the Poisson tables by using the $c = 0$ column and equating λT with np or by the use of a modern hand calculator. Since this law is just the first term of the Poisson distribution (the probability of no failures for a stated time), the probability of the system failing (P_f), then, can be found by summing all of the remaining terms. Since this is a tedious process, the complementary law can be used. The equation is

$$P_f = 1 - P_s \quad \text{or} \quad P_f = 1 - P_0 \quad \text{or} \quad P_f = 1 - R$$

The three reliability questions of the previous section can now be solved for the exponential case. Question 1 can be solved by using the formula, or the tables, directly. Questions 2 and 3 must be solved by the inverse application of the formula or tables. It is easier to solve the formula inversely first and then use the new formulas to solve the questions.

$$R = \exp(-T/\overline{m})$$
Take the reciprocal of both sides: $1/R = \exp(T/\overline{m})$
Take the natural log of both sides: $\ln(1/R) = T/\overline{m}$
Solve for T and $\overline{m}$.

The three formulas, then, are:

$$R = \exp(-T/\overline{m})$$

$$T = \overline{m} \ln(1/R)$$

$$\overline{m} = T/\ln(1/R)$$

These formulas are the same whether time-terminated of failure-terminated data are used.

Reliability using the Exponential and Time-Terminated Data

Example 6.1 (*This is an example of reliability question 1 for time-terminated data*) Four hundred units are tested for 10 hr and are not replaced when they fail. At the end of the test there are 40 failures.

What are the 10-hr, 100-hr, and 1,000-hr reliabilities for the unit? What is the 100-hr probability of failure?

Solution:

1. $\bar{s} = \dfrac{400 + 360}{2} = \dfrac{760}{2} = 380$ units
2. $\lambda = f/\bar{s}t = 40/380(10) = 0.0105$ fph
3. 10-hr reliability $= R = e^{-\lambda T} = e^{-0.0105(10)} = 0.900$ or 90%
4. 100-hr reliability $= R = e^{-0.0105(100)} = 0.349$ or 34.9%
5. 1000-hr reliability $= R = e^{-0.0105(1000)} = 0.00003$ or 0.003%
6. P_f(for 100 hr) $= 1 - R = 1 - 0.349 = 0.651$

Notes on Example 6.1:

1. The probability of no failures (P_o or P_s or R) reduces as the time frame (T) increases.
2. Total test time was only estimated (380×10) at 3800 hr based on average survival rates. If actual failure times had been available, they would have been used instead of $\bar{s}$.
3. If failed units had been repaired and replaced, the total test time would have been 4000 (400×10) and R would have been 0.368 (for 100 hr).
4. The mean life is 95 hr (3800/40). The alternate formula using mean life would have given the same results ($e^{-\lambda T} = e^{-T/\bar{m}}$).
5. The above example can be interpreted as follows (using 100-hr reliability as an example): We are 35% certain that any one unit can operate for 100 hr without failure *or*, out of 100 units, 35 will operate for 100 hr without failure, while 65 will not (on the average).

Example 6.2 Suppose that the failure rate of Example 6.1 was improved from 40 failures to 20. What is the 100-hr reliability?

Solution:

1. $\bar{s} = \dfrac{400 + 380}{2} = \dfrac{780}{2} = 390$ units
2. $\lambda = 20/390(10) = 20/3900 = 0.00513$ fph
3. $R = e^{-0.00513(100)} = 0.5987$ or 59.9%

Note that, as the failure rate λ is decreased, the reliability is increased. Since the mean life is the reciprocal of the failure rate, just the opposite relation exists between mean life and reliability. As the mean life (m or $\bar{m}$) is increased, reliability is increased.

Example 6.3 (*This is an example of reliability question 2 for time-terminated data*) Using the data from Example 6.1, what must the mission time (T) be for a desired reliability of (1) 90%? (2) 95%? (3) 80%?

Solution:

(1) $R = 90\%$
$T = \overline{m}[\ln(1/R)]$
$T = (1/0.0105)[\ln(1/0.90)] = 95(0.10536) = 10.01$

Note how this answer matches exactly with the answer obtained in Example 6.1 ($T = 10$ and $R = 90\%$).

(2) $R = 95\%$
$T = 95[\ln(1/0.95)] = 4.873$ hr
(3) $R = 80\%$
$T = 95[\ln(1/0.80)] = 21.2$ hr

Note how the mission time increases (can be specified higher) as the desired reliability is decreased. These relationships (T to R to $\overline{m}$) remain the same for all levels of activity and for all types of distributions.

Example 6.4 If the mean life is 95 hr, how many failure-free hours can be expected at a reliability of (1) 95%? (2) 80%?

Solution:

(1) $R = 95\%$
$T = \overline{m} \ln(1/R) = 95 \ln(1/0.95) = 4.87$ hr
(2) $R = 80\%$
$T = 95 \ln(1/0.80) = 21.2$ hr

Note: This example can be interpreted to mean 95 of 100 systems can be expected to operate for 4.87 hr without a failure, but only 80 of 100 can be expected to operate for 21.2 hr without a failure.

Example 6.5 (*This is an example of reliability question 3 for time-terminated data*) What mean life must be specified for: (1) A desired reliability of 90% and a desired mission time of 10 hr? (2) A desired reliability of 80% and a desired mission time of 10 hr? (3) A desired reliability of 90% and a desired mission time of 100 hr? (4) A desired reliability of 90% and a desired mission time of 1 hr?

Solution:

(1) $R = 90\%$ and $T = 10$ hr
$\overline{m} = T/[\ln(1/R)]$
$\overline{m} = 10/[\ln(1/0.90)] = 10/(0.10536) = 94.91$ hr

Note how this matches almost exactly with Examples 6.1 and 6.3.

(2) $R = 80\%$ and $T = 10$ hr
$\overline{m} = 10/[\ln(1/0.80)] = 10/(0.22314) = 44.8$ hr
(3) $R = 90\%$ and $T = 100$ hr
$\overline{m} = 100/[\ln(1/0.90)] = 949$ hr
(4) $R = 90\%$ and $T = 1$ hr
$\overline{m} = 1/[\ln(1/0.90)] = 9.49$ hr

Note the interrelationships of T, R, and $\overline{m}$ (and that these relationships remain the same as explained earlier).

Reliability using the Exponential and Failure-Terminated Data

In Example 6.6, failure-terminated data are used. Example 6.6 will be used with all distributions in this chapter and in Chapter 8 to provide a basis of comparison. It must be emphasized that this example does not exactly fit any of the distributions used (except for the Weibull). Because of this, some of the answers to some of the distributions may seem ridiculous at times (and may well be).

Example 6.6 (*This is an example of reliability question 1 for failure-terminated data*) Six units are tested to failure with the following failure times: 53, 66, 80, 82, 89, and 91 hr. Find the reliability for a mission of 40 hr.

Solution:

$$\overline{m} = (53 + 66 + 80 + 82 + 89 + 91)/6 = 461/6 = 76.8 \text{ hr}$$

$$R = \exp(-T/\overline{m}) = \exp(-40/76.8) = 59.4\%$$

Interpretation: 59.4% of the items (those with a mean life of 76.8 hr) will last longer than 40 hr; *OR* there is a 59.4% probability that any one item will last longer than 40 hr (these are equivalent statements, they mean the same thing).

Example 6.7 (*This is an example of reliability question 2 for failure-terminated data*) Using the data of Example 6.6, find the mission time for a reliability of 90%.

Solution:

$$T = \overline{m}[\ln(1/R)] = 76.8[\ln(1/0.90)] = 8.09 \text{ hr}$$

Interpretation: 90% of the items will last longer than 8.09 hr.

Example 6.8 (*This is an example of reliability question 3 for failure-terminated data*) What mean life must be specified for Example 6.6 in order to have a reliability of 90% and a mission time of 40 hr?

Solution:

$$\overline{m} = T/[\ln(1/R)] = 40/[\ln(1/0.90)] = 380 \text{ hr}$$

Interpretation: In order to have 90% of the items last longer than 40 hr, the product must be produced to an average mean life of 380 hr (assuming the failures are exponentially distributed).

6.4 The Binomial Distribution

The binomial formula is

$$P(d) = C_d^n p^d (1 - p)^{n-d}$$

Since reliability is the probability of zero failures, the d in the preceding formula becomes zero and the C reduces to one. Therefore, the reduced formula becomes $P(d) = (1 - p)^n$. For reliability, $p = \lambda$ (the failure rate) and $n = T$ (the mission time). Therefore, the binomial formula for reliability is

$$R = (1 - \lambda)^T \quad \text{or} \quad R = (1 - 1/\overline{m})^T$$

In order to answer reliability questions 2 and 3, the preceding formula must be solved for $\overline{m}$ and for T. The three formulas, then, are

$$\left.\begin{array}{l} R = (1 - 1/\overline{m})^T \\[2mm] T = \ln R / \ln[1 - (1/\overline{m})] \\[2mm] \overline{m} = 1/\{1 - \exp[(\ln R)/T)]\} \end{array}\right\}$$

Note: It is important here that the failure rate be less than one (<1.00). If it is not, the time units must be changed to the next lowest increment. For instance, suppose that 10 units with 10 failures had been tested for $1/2$ hr each. The failure rate is $10/10(0.5) = 2.00$ failures/ hr. If the time units are changed to minutes, the failure rate then becomes $10/10(30) = 0.03333$ failures/min. Of course, the mission time must also be changed to minutes; the time units must always match or the answers are meaningless.

The three reliability questions will now be solved using the binomial distribution and the previous three failure-terminated data examples (Examples 6.6, 6.7, and 6.8) for comparison. The three previous discrete examples can also be solved using the binomial, but this will be left to the student as an exercise. The binomial answers will be almost identical to those already determined by use of the exponential.

Example 6.9 (*This is an example of reliability question 1 for failure-terminated data*) Solve Example 6.6 using the binomial distribution.

Solution:

$$R = (1 - 1/\overline{m})^T = (1 - 1/76.8)^{40} = 59.2\%$$

Example 6.10 (*This is an example of reliability question 2 for failure-terminated data*) Solve Example 6.7 using the binomial distribution.

Solution:

$$T = \ln R/\ln(1 - 1/\overline{m}) = \ln(0.90)/\ln(1 - 1/76.8)$$

$$= -0.10536/\ln(0.987) = -0.10536/-0.0131 = 8.04 \text{ hr}$$

Example 6.11 (*This is an example of reliability question 3 for failure-terminated data*) Solve Example 6.8 using the binomial distribution.

Solution:

$$\overline{m} = 1/\{1 - \exp[(\ln R)/T]\} = 1/\{1 - \exp[\ln (0.90)/40]\}$$

$$= 1/[1 - \exp(-0.10536/40)] = 1/[1 - \exp(-0.002634)]$$

$$= 1/[1 - 0.99737] = 1/0.00263 = 380$$

Note how the answers using the binomial are almost identical to those using the exponential. This will almost always be the case.

6.5 The Normal Curve

As has already been stated, the normal curve is a continuous distribution and must use failure-terminated data. In the case of reliability, this means that the units must all be tested to failure, or that the mean and standard deviation must be assumed (by management and/or engineering) as if they had been derived from units tested to failure. These kinds of assumptions come from (1) educated guess, (2) comparison to similar products run on similar processes, and/or (3) on-going production information about the product from a stable process where the product has been produced for a long period of time.

In calculating reliability using the normal distribution, the probability of survival, or the probability of zero failures, is the area under the curve above the specified time. Therefore, reliability is $1 - P(z)$ (see Chapter 2). The mission time is equivalent to the X in the z formula of Chapter 2 and must be determined in exactly the same way that X was determined in Chapter 2. Mean life, in reliability, is equivalent to the mean in Chapter 2 and must be determined exactly the way the mean was determined in Chapter 2.

The formulas are

$$z = (T - \overline{m})/\sigma \text{ (if sample data are used, } \sigma = s)$$

$$R = 1 - P(z) \text{ or } 1 - P_z \text{ or } 1 - \alpha$$

(where α is the area under the curve below the specified value — α is obtained from the body of Table A1).

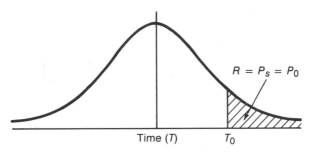

Fig. 6.2 Normal curve reliability.

Example 6.12 (*This is an example of reliability question 1 for failure-terminated data*) Solve Example 6.6 using the normal distribution; assuming a mean of 76.8 and a standard deviation of 14.6 hr.

Solution:

$$z = (T - \overline{m})/s = (40 - 76.8)/14.6 = -2.52$$

$$P(z) = \alpha = 0.0059 \text{ (Table A1)}$$

$$R = 1 - P(z) = 1 - 0.0059 = 99.41\%$$

Example 6.13 (*This is an example of reliability question 2 for failure-terminated data*) Solve Example 6.7 using the normal distribution.

Solution:

$$z_{1-0.90} = z_{0.10} = (T - m)/s = -1.282 = (T - 76.8)/14.6$$

$$T = 76.8 - 1.282(14.6) = 58.1 \text{ hr}$$

Example 6.14 (*This is an example of reliability question 3 for failure-terminated data*) Solve Example 6.8 using the normal distribution.

Solution:

$$z_{1-0.90} = z_{0.10} = (T - \overline{m})s = -1.282 = (40 - \overline{m})/14.6$$

$$\overline{m} = 40 + 1.282(14.6) = 58.7 \text{ hr}$$

Note that the answers using the normal are quite different from the answers determined by using the two discrete distributions. The decision as to type of distribution is obviously critical. Unfortunately, this decision is not easily arrived at and usually requires some high-level statistical analysis to determine. Fortunately, the Weibull distribution does not require this decision because of the wide range of distribution types automatically contained in its calculations and formulas. This is one reason that the Weibull has become so popular.

6.6 The Weibull Distribution

In the past few years the Weibull distribution has become increasingly favored by reliability engineers. Weibull distributions have three parameters — scale (G), shape (B), and location — and can be made to take on any desired form by simply changing one or more of these three values (Fig. 6.3). The scale parameter G is also called the characteristic life and the shape B is also called the Weibull slope.

In reliability work, the location parameter is usually set at zero and is therefore ignored in the formulas. Setting this parameter at zero is tantamount to saying that the distribution of failure times started at zero time (the start of the test). This condition shows up as a straight line on the Weibull graph. A curved line on the graph indicates that the distribution of failures started sometime other than time zero, which is a difficult concept to imagine in real life. A problem of this nature (a curved line), requires expert assistance from an experienced statistician and is beyond the scope of this text.

The shape parameter (B) controls the shape of the resulting curve, so that almost any type of curve can be generated by changing this value. On the Weibull graph, the shape parameter is indicated by the slope of the straight line. A shape parameter of 1 (B = 1) indicates

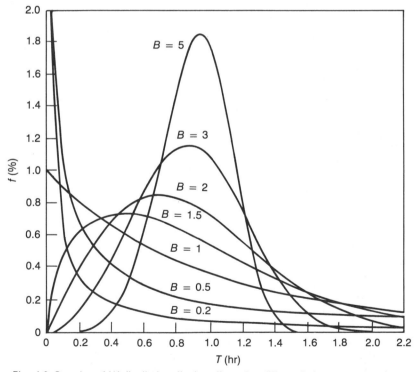

Fig. 6.3 Graphs of Weibull density functions for different slope parameters.

an exponential distribution (see Weibull formula 1, following). The normal is shown by a shape parameter (slope) of about 3.4 (actually the range between 2.7 and 3.7 is usually considered normal).

The scale parameter (G) locates the line on the graph, in much the same way as the mean does for the normal, and performs the same function in the Weibull as does the mean in the normal. However, the scale parameter G is *not* the same thing as the mean life. Mean life is determined by a special formula. This is because mean life in the Weibull changes with changing life requirements (mission time, T). It is obvious from Fig. 6.3 that a different mission time T will move the distribution to a different location on the time scale, which will change the shape of the curve, requiring a different mean life.

It is apparent that the Weibull distribution can be used in place of the exponential or the normal, and this is part of the reason for the Weibull's wide acceptance among reliability engineers. The fact that it can be made to assume a very large variety of shapes, and therefore be able to cover many varied types of failure rate functions, is another reason for its attractiveness. And finally, it is not required, in using the Weibull, to determine the necessary distribution before analysis. The Weibull is supposed to do that automatically.

The Weibull, as applied to reliability, is considerably more complex than either the exponential or the normal. For one thing, the mean time to failure has two parameters rather than one, and cannot be obtained directly as in the exponential or normal. Also when calculating R, the shape parameter must be determined first.

Weibull distributions, where B is less than 1.0, are used to analyze early failures (the break-in stage, also called infant mortality), while those where B is larger than 1.0 are used to analyze wear-out failures. Of course, where B is exactly equal to 1.0, the distribution becomes exponential (as previously stated), and this is used to analyze failures that occur at a constant rate (the operating stage as shown on Fig. 1.1).

The Weibull reliability formulas are:

1. $R = \exp[-(T/G)^B]$

    where

    R = reliability

    T = mission time

    B = the Weibull slope (the shape parameter)

    G = the characteristic life (equivalent to the mean of the normal, but NOT the same)

2. $T = G(-\ln R)^{1/B}$

3. $\overline{m}$ must be calculated from a given R and T:

$$\overline{m} = \exp\{\ln T - [\ln \ln(1/R)]/B\}$$

4. The shape parameter B (also called the Weibull slope) must be determined. This can be done either by reading it from the graph, when the graph is available, or by solving the following least-squares formula:

$$B = \frac{\{\Sigma(\ln X)\ln \ln[1/(1 - Y)]\} - (\Sigma(\ln X) \Sigma\{\ln \ln[1/(1 - Y)]\})/n}{\Sigma(\ln X)^2 - (\Sigma \ln X)^2/n}$$

where

X = the times to failure

Y = the percentiles from the median (50%) rank of Table A5

Note: When the Y values, the percentile values from Table A5, are used in these formulas, the decimal fraction form must be used (0.11 instead of 11%). Median (50%) ratios (not percentiles) can also be determined from the following formulas (from Johnson, *The Statistical Treatment of Fatigue Experiments,* Elsevier, New York, 1964):

(a) For sample sizes n of 20 or less.

$$Y = 1 - 2^{(-1/n)} + [(j - 1)/(n - 1)][2^{(1-1/n)} - 1]$$

(b) For $n > 20$.

$$Y = [j - 0.30685 - 0.3863 (j - 1)/(n - 1)]/n$$

where

Y = MR = median (50%) rank ratios (percentiles are determined by multiplying the ratios by 100)

j = failure ranking (j = 1 for the first failure, j = 2 for the second, etc.)

n = sample size (number of units tested, not the time sample T).

Although these formulas are only approximations, they are very close and more than adequate for the formulas and procedures in this text. Tables of more precise ratios (and including percentile ranges from 0.5% to 99.5%), derived from basic theory using the z and F tables, are available (Hald, *Statistical Tables and Formulas,* Wiley, New York, 1952).

5. The Y intercept for the line of best fit is

$$a = \{\Sigma \ln \ln[1/(1 - Y)]\}/n - B[\Sigma (\ln X)/n]$$

6. The scale parameter G is also the Weibull characteristic life and is used in the Weibull formulas in the same way that mean life is used in the normal. G can be either read from the graph, if a graph is available, or calculated from the following formula.

$$G = \exp[(-0.00033 - a)/B]$$

The constant 0.00033 is derived from $\ln \ln[1/(1 - 0.632)]$.

Mean life, in the Weibull, is a function of the mission time (see Fig. 6.3) and so is different for different mission times. This is not true of other distributions where mean life is constant and does not change with changes in the mission time.

There are two ways to solve for Weibull reliability: graph or formula. The graphical method is relatively easy, but is limited to solving the first two reliability questions only. The formula method, on the other hand, can solve all three types of the reliability question, but the calculations are rather complex and exacting. One method of simplifying the formula calculations is to determine the Weibull parameters (B and G) from the graph first and then to use them in the formulas to solve all three problems. (The determination of B and G is the most difficult of all the Weibull calculations.) However, precise answers can be derived only from the formulas. A computer can be extremely useful here, and computer programs are available for making these calculations (even a scientific hand calculator is difficult to use for calculating B and G).

The Weibull test procedures are:

1. Determine the sample size. The following conformance table can assist in determining minimum sample sizes:

Conformance	Sample Size
95%	6
98%	12
99%	20

2. Test to failure or until each part exceeds three times the performance requirements. This limit (three times the performance requirements) is an arbitrary one imposed by management and/or engineering judgment, and can be increased, depending on the criticality of the product. It probably should not be decreased.
3. List the failure times from low to high.
4. List the median (50%) rank percentiles from Table A5 (or by using one of the formulas), from low to high. If the test was terminated before all samples had failed, list only the number of percentages that match the number of failed

Handwritten margin notes:

6.19

T	%
450	3.4
510	8
555	13
585	18
610	23
625	28
655	33
675	32
690	43
710	48
730	52
750	57
765	62
785	67
800	72
825	77
850	82
885	87
915	92
1000	97

$T \approx 400$

$R = 1 - .08 = .92$

$T = 500, R = .95;$

$m = \exp[\ln 500 - (\ln \ldots)]$

$= 885$

units, from the beginning of the table. For instance, if only 8 had failed from a sample of 12, only the first 8 values from the 12 table would be used.

5. Plot the points from each pair of values, with failure times on the X axis and percentages on the Y axis.

6. Fit a straight line to the plotted points. If the line is obviously curved, the procedures presented in this text will not work. A curve means that the time frame does not start at zero, a situation that is not supposed to occur in reliability. The solution to this problem is beyond the scope of this text (see your friendly statistician).

7. Read the reliability for a stated mission time from the graph. Enter the graph on the X axis at the stated mission time, move up to the line just fitted, and then read the failure time from the Y axis (the Y axis represents failure percentages, not reliabilities). Reliability then is one minus the Y value ($R = 1 - Y$).

8. Read the mission time for a stated reliability from the graph. Enter the graph on the Y axis at $1 - R$ (remember that the Y axis represents failure percentages, not reliabilities), move to the right to the line just fitted, and read the mission time from the X axis.

9. The mean life, or new mean, cannot be determined from the graph.

10. If the formulas are to be used instead of the graph, omit steps 5 through 9 and use the following steps.
   a. Calculate the parameters B and G from the least-squares formulas (or construct the graph and read them from the graph, if desired).
   b. Calculate R, T, and $\overline{m}$ from the formulas.

Example 6.15 (*This is an example of reliability questions 1, 2, and 3 for failure-terminated data*). Solve Example 6.6 using the Weibull graph. Assume that the sample size was six and that all units failed.

Solution:

Step 1. List the pairs of values.

Failure Times	Median — 50% (from Table A5) n = 6
53	11
66	26
80	42
82	58
89	74
91	89

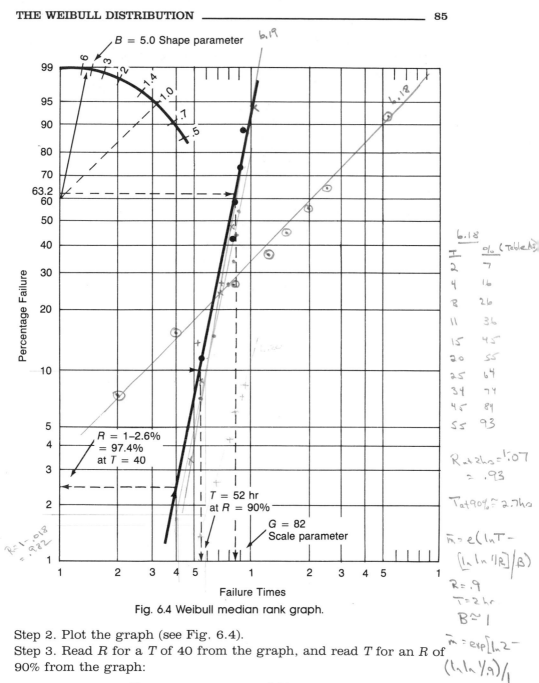

Fig. 6.4 Weibull median rank graph.

Step 2. Plot the graph (see Fig. 6.4).
Step 3. Read R for a T of 40 from the graph, and read T for an R of 90% from the graph:

$$R = 1-1.6\% = 99.4\% \text{ at } T = 40$$

$$T = 52 \text{ hr at } 1 - R \quad (1 - 0.90)$$

Step 4. Read the slope, $B = 5$, from the graph by drawing a line parallel to the line of best fit through the origin (small circle) of the Weibull slope exhibit at the upper left corner of the graph. The slope B is the intersection of this parallel line with the Weibull arc located in the upper left corner of the graph.

Step 5. Read the scale parameter, $G = 82$, from the X axis at the bottom of the graph. Enter the graph on the Y axis at 63.2%, move right to the line of best fit (not the parallel line used to determine B), and read G from the X axis at the bottom of the graph.

Note: These values do not exactly match the same values determined by the formulas (Example 6.15). This is because a line fitted by sight is seldom accurate.

Example 6.16 Solve Example 6.15 by using the Weibull formulas.

Solution:

1. Solve for the Weibull slope, B:

$$B = \frac{\{\Sigma(\ln X)\ln \ln[1/(1 - Y)]\} - (\Sigma\{\ln X\} \Sigma(\ln \ln[1/(1 - Y)]\})/n}{\Sigma(\ln X)^2 - (\Sigma \ln X)^2/n}$$

$$B = \frac{-11.94275 - [25.9482(-3.009786)]/6}{112.43347 - (25.9482)^2/6} = 4.987.$$

2. Solve for the intercept, a:
$a = \{\Sigma \ln \ln[1/(1 - Y)]\}/n - B[\Sigma(\ln X)/n]$
$a = -3.009786/6 - 4.987(25.9482)/6 = -22.0689.$

3. Solve for the characteristic life, G:
$G = \exp[(-0.00033 - a)/B]$
$G = \exp[(-0.00033 - (-22.0689))/4.987] = 83.51.$

4. $R = \exp[-(T/G)^B]$
$R = \exp[-(40/83.51)^{4.987}] = 0.975$ or 97.5%.

5. $T = G(- \ln R)^{1/B} = 83.51(- \ln 0.90)^{1/4.987} = 53.2$ hr

6. $\overline{m} = \exp\{\ln T - [\ln \ln(1/R)]/B\}$
$\overline{m} = \exp\{\ln 40 - [\ln \ln(1/0.90)]/4.987\} = 62.8.$

Example 6.17 Solve Example 6.15, using the Weibull formulas and assuming that the original sample size was 10; there were six failures out of 10 units (use $n = 6$ for all calculations).

Solution:

Step 1. List the pairs of values.

Failure Times	Median — 50% (Table A5) $n = 10$
53	7
66	16
80	26
82	36
89	45
91	55

Step 2. Solve for R, T, and $\overline{m}$.

$$R = 98.1\%$$

$$T = 59.3 \text{ hr}$$

$$\overline{m} = 70.0 \text{ hr}$$

Note the difference between the preceding answers and Examples 6.15 and 6.16. The proof of these answers is left to the reader.

It should be noted that the new mean life value can be determined without recourse to sample information, just as in all other distributions. In order to do this with the Weibull, however, the slope B and scale parameter G must be assumed (just as the mean life and standard deviation are assumed in the normal, and for the same reasons). In the preceding problem, the values already calculated from the sample were used, but this is not necessary. Any B and G values can be assumed and used in solving for the new mean life, and thus provide valuable design and contract analysis information even before the design begins.

6.7 Comparing the Distributions

Throughout most of this chapter, only one failure-terminated example was used for each of the different distributions (Example 6.6). This was done so that the different answers could be compared. Table 6.1 summarizes these answers for convenient study. The binomial distribution was not included because its answers are identical to those of the exponential.

The following four examples are presented as a comparison to the previous problem. The first is exponentially distributed (approximately) and the second is approximately normally distributed. The third and fourth are recalculations of Example 6.6 using a different reliability in one case and a different mission time in the other. Only the answers will be given. The proof of the answers will be left to the reader as an exercise.

Example 6.18 A sample of 10 units was tested to failure. The failure times, in hours, were: 2, 4, 8, 11, 15, 20, 25, 34, 45, and 55. Find the reliability for a stated mission time of 2 hr. Find the mission time for

Table 6.1 Comparison of Reliability Values

Distribution	R (%)	T (hr)	$\overline{m}$ (hr)
1. Exponential	59.4	8.09	380
2. Normal	99.4	58.10	58.7
3. Weibull	97.5	53.20	62.8

$\exp: \quad R = e^{-\lambda T}$

$\overline{m} = \dfrac{309}{10} = 21.9 \text{ hrs}$

$\lambda = 0.0457$

$R = e^{-.0457(2)} = 0.913$

$.9 = e^{-.0457\,T}$

$-.1054 = -.0457T$

$T = 2.34 \text{ hrs},$

$.9 = e^{-\frac{T}{m}\,2}$

$-.1054 = -\frac{1}{m}\,2$

$m = 18.97$

normal

a stated reliability of 90%. Find the new mean needed for a reliability of 90% and a mission of 2 hr.

Solution:

Distribution	R (%)	T (hr)	$\overline{m}$ (hr)
1. Exponential	91.3	2.31	19.0
2. Normal	86.3	−0.95	24.9
3. Weibull	92.7	2.76	17.2

Example 6.19 A sample of 20 units was tested to failure. The failure times, in cycles per use, were: 450, 510, 555, 585, 610, 625, 655, 675, 690, 710, 730, 750, 765, 785, 800, 825, 850, 885, 915, and 1000. Find the reliability for a mission time of 500 cycles. Find the mission time for a reliability of 95%. Find the new mean life for a reliability of 95% and a mission time of 500 cycles.

$\overline{m} = \dfrac{14370}{20} = 719 \quad R = e^{-500/719} = 0.50$

$.95 = e^{-T/719}$

Solution:

$.95 = e^{-500/\overline{m}}$

$-.0513 = -500/\overline{m}$

$\overline{m} = 9746$

$-.0513 = -\dfrac{T}{719}$

$T = 37$

Distribution	R (%)	T (cycles)	$\overline{m}$ (cycles)
1. Exponential	49.9	36.9	9,748
2. Normal	93.9	488	732
3. Weibull	92.6	467	837

Example 6.20 Solve Example 6.6 for a desired reliability of 99% (instead of the 90% used before) and a mission time of 40 hr.

Solution:

T	%
53	11
66	86
80	47
82	58
89	74
91	89

Distribution	R (%)	T (hr)	$\overline{m}$ (hr)
1. Exponential	59.4	0.77	3,980
2. Normal	99.4	42.77	74
3. Weibull	97.5	33.20	101

Example 6.21 Solve Example 6.6 for a desired reliability of 90% and a mission time of 50 hr (instead of the 40 hr used before).

Solution:

Distribution	R (%)	T (hr)	$\overline{m}$ (hr)
1. Exponential	52.2	8.09	475
2. Normal	96.7	58.10	68.7
3. Weibull	92.5	53.19	78.6

Analysis of the Comparisons

An examination of the preceding problems, plus the relationships of reliability, mission time, and mean life, provide some useful tools for analysis. The conclusions are summarized below.

1. As the mission time (T) increases, the reliability decreases. In other words, the longer you want something to last, the less likely it will do so. This is exactly the relationship noted in real life.

2. As mean life is increased, reliability is increased. The longer a product will last on the average, the more we can expect the product to last for the desired mission time.

3. As failure rate is increased, reliability decreases. This follows from the inverse relationship of failure rate to mean life ($\lambda = 1/\overline{m}$).

4. As reliability is increased, mean life must be increased or mission time must be decreased, or combinations of both. If a higher reliability is desired, something must be done. The easiest change to make is to decrease the product requirements; decrease the mission time or the reliability specifications or both. This may not, and usually is not, possible. The customer may refuse. Therefore, the mean life must usually be increased. This can be done in one of two ways, or combinations of both. The first is to redesign and reduce the failure rate limit—the design limit (lower the bottom of the bathtub curve, Fig. 1.1). The second would be to redesign the process to increase the product quality. Either can be an expensive proposition. The decision as to which one and/or how much of each is a high-level management problem.

5. The Weibull usually provides a close approximation of the normal when the distribution is normal and to the exponential when the distribution is exponential. The distribution of failure times for the problem of Table 6.1 (see Example 6.6) is neither exponential nor normal, although it comes much closer to normal than it does to exponential. Note how the Weibull closely approximates the normal in this table, and how much the exponential differs from both. Such is also the case in Example 6.19, which is a normal distribution. However, in Example 6.18 the failure times are exponentially distributed and so the Weibull closely approximates the exponential in this case.

6. It should be noted that the Weibull does not always approximate the distributions, as was explained in point 5. It does approximate a wide normal curve (with weak central tendencies) quite well, but does not do nearly as well representing a narrow curve (with strong central tendencies). For instance, the normal distribution of failure times in Example 6.19 when plotted on the Weibull graph shows a Weibull slope (B) of 5. Since the Weibull only approximates the normal when its slope is in the range between 2.4 to 3.7, the Weibull in this case is showing an incorrect slope. Note that the Weibull answers, however, still closely approximate the normal, as they are suppose to.

7. If the distribution is normal, the exponential solution values will be lower, usually quite lower, than the normal and the Weibull solution will roughly match the normal. If the distribution is exponential, the normal solution will be lower than the exponential and

the Weibull will roughly match the exponential. There will be differences, at times, but they will usually be small unless higher levels of confidence are used (see Chapter 8).

Testing Goodness of Fit

Since determining the right distribution is so important to reliability, some means of making the proper choice is needed. In reliability, one of the best methods of doing this is with probability plots (a form of "goodness-of-fit" analysis). Probability plots involve the use of probability paper. This type of paper is illustrated in Fig. 6.5 (for exponential data) and 6.6 (for normal data).

Figure 6.5 represents a semilogarithmic exponential paper with the failure times plotted on the x axis and the reliability percentages on the y axis. The reliability percentages are determined by subtracting the median rank values of Table A5, for the particular sample size, from 1.00. The median rank (50%) values are matched to the failure values in ascending order, exactly as in the Weibull-graph solution method. Then each median rank value is subtracted from 1.00 to find the reliability values. Each pair of values (failure time/reliability) is then plotted and a line is fitted to the points. If a straight line cannot be fitted, the distribution is not exponential. The line will descend from left to right.

Example 6.22 Plot Example 6.18 on exponential paper. The table below gives the point pairs for the graph and Fig. 6.5 gives the answers.

Failure Times	Median (50%) Rank (from Table A5)	1 − 50% Rank
2	7	93
4	16	84
8	26	74
11	36	64
15	45	55
20	55	45
25	65	35
34	74	26
45	84	16
55	93	7

Figure 6.6 represents a plotting paper for a normal distribution. The median rank (50%) values from Table A5 are matched to the failure times in ascending order, just as in the Weibull. Then the pairs of values are plotted onto the normal probability paper (the median

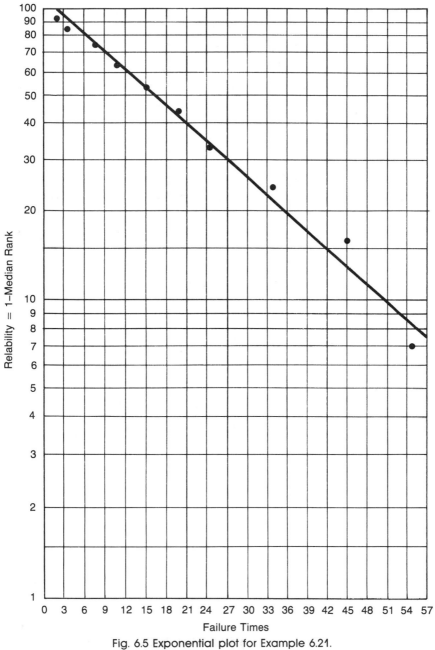

Fig. 6.5 Exponential plot for Example 6.21.

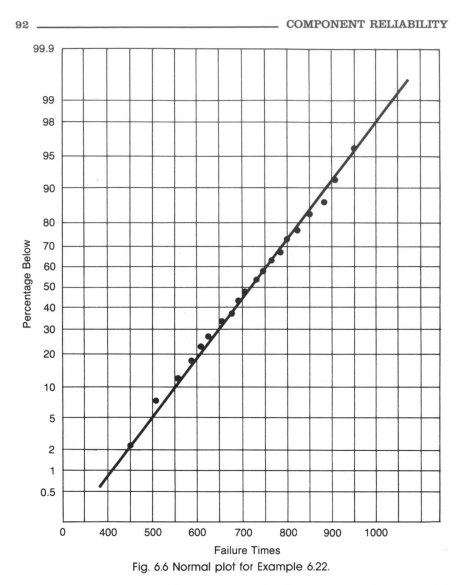

Fig. 6.6 Normal plot for Example 6.22.

ranks are *not* subtracted from 1.00). If the line of best fit is a straight line, the distribution is normal.

Example 6.23 Plot Example 6.19 on normal probability paper. Figure 6.6 gives the solution. Development of the table of point pairs is left to the reader as an exercise.

Problem Interpretation

Most of the examples in this chapter assume three different types of inputs; a desired reliability (R), a desired mission time (T), and data for calculating average mean life ($\overline{m}$). From these three types of inputs, the answers to the three reliability questions can be determined. In actual practice, however, only two of the three types of inputs will usually be available at any one time (or for any one problem). The two types of available input, at that time, would then be used to calculate the missing data (to answer the reliability question relating to the missing data). The reason for the presentation procedure used in this text can be illustrated using the following scenario.

Suppose a prospective order is received for the product described in Example 6.19, and that the order requires that the product be 95% reliable. The customer then asks, "What mission time should be specified for your product at a reliability of 95%?" An average mean life is then determined (either by assumption or by actual test), a distribution is chosen, and a mission time is calculated.

Let us suppose that the exponential is chosen and that the calculated mission time is 36.9 hr (Example 6.19). Let us further suppose that management has information that suggests that this mission time will be unacceptable to the customer. Management then might ask the reliability analyst a series of questions designed to assist management in the devising of an optimum sales/production strategy.

For instance, management might want to know what the reliability would be for a given, or assumed, mission time (reliability question 1). Perhaps it might be possible to convince the customer to reduce his or her reliability requirements or accept a lower than desired mission time or both. If not, the product must be improved (which usually means spending a lot of money on a new product design or process design or both). Or the order could be refused (which may well reduce future, as well as present, revenue). At this time, of course, management will want to know what the new production mean life must be in order to meet the expected reliability and mission time requirements (reliability question 3), or what it must be in order to meet requirements it hopes to be able to sell to the customer.

A similar series of analysis questions and calculations must be used if the customer requirements list only the mission time, or if both mission and reliability are specified. The analyst can be well prepared for this searching analysis if he or she assumes a realistic mission time or reliability (depending on which has not been specified) and completes the calculations prior to management's contract evaluation (has the answers ready before the questions are asked). In this text, this procedure is followed in most examples (all three reliability questions are answered in the same problem). This is not only true in this chapter, but also for confidence limits in Chapter 8 (where there are four confidence limit questions) and for maintainability in Chapter 9.

6.8 Conditional Reliability

Sometimes it is necessary to know what the probability would be that a system will last for a stated length of time beyond a given minimum time length. The formula is

$$R(T_{0,t}) = \frac{R(T_0 + t)}{R(T_0)}$$

where

$R(T_0 + t)$ = probability that a system or part will operate for $T_0 + t$ hr without a failure

$t = T_1 - T_0$ hr or the change in mission hours from T_0 to T_1

When the times to failure are exponentially distributed, conditional reliability is just a simple exponential calculation, as follows:

$$\frac{e^{-\lambda(T+t)}}{e^{-\lambda T}} = e^{-\lambda t}$$

However, this simple relationship works only with the exponential distribution. For all other distributions, Eq. (6.1) must be used as given.

Example 6.24 A certain machine has been in operation for 1200 hr without preventive maintenance. The next two days (48 hr) are critical in the production cycle. Management decides that time will be allowed to perform preventive maintenance (shut down the machine, oil, inspect, etc.) if the calculated reliability for the next 48 hr is less than 95%. Calculate R for the machine under two assumptions:

1. The times to failure are exponentially distributed with a mean time to failure of 1300 hr.
2. The times to failure are normally distributed with a mean time to failure of 1300 hr and a standard deviation of 100 hr.

Solution:

1. $\dfrac{e^{-(T+t)/m}}{e^{-T/m}} = \dfrac{e^{-(1200+48)/1300}}{e^{-1200/1300}} = \dfrac{0.3829}{0.3973} = 0.9638$

or

$$e^{-t/m} = e^{-48/1300} = 0.9638$$

2. $\dfrac{R(1200 + 48)}{R(1200)} = \dfrac{0.6985}{0.8413} = 0.8303$

Note: 0.6985 is derived as follows:

$$z = (1248 - 1300)/100 = -0.52$$

$$P(z \leqslant -0.52) = 0.3015 \text{ (Table A1)}$$

$$R_{(1200+48)} = 1 - 0.3015 = 0.6985$$

[0.8413 is similarly derived].

Terms and Definitions

1. **Failure analysis.** The use of probability distributions to predict the probability of failures within certain given time limits.
2. **Hazard rate.** The instantaneous failure rate — the failure rate at any instant of time. In most reliability systems the hazard rate and the constant failure rate (mature system) are assumed to be equal.
3. **Ratio failure rate.** Percentage of failure or failures per part.
4. **Exponential failure law.** The first term of the Poisson expansion (probability of zero failures) and a special case of the exponential distribution: $P_s = P_0 = R = e^{-\lambda T}$.
5. **Probability of survival.** Probability of zero failures within the designated system life: $P_s = e^{-\lambda T}$.
6. **Probability of failure.** $P_f = 1 - e^{-\lambda T}$. The sum of all the Poisson terms except the first.
7. **Mean life.** Average life of the component or system.
8. **100-hour reliability.** Probability of a system surviving (zero failures) for 100 hr.
9. **MTBF.** Mean time between failures — equivalent to m or $\overline{m}$.
10. **MTTF.** Mean time to failure (to first failure) — equivalent to m or $\overline{m}$; usually considered the same as MTBF, but not always the case.

Practice Problems

1. The failure rate for a hand calculator is 0.02 fph. (1) Calculate the mean time between failures. (2) What is the probability of its failing within 4 hr? (3) What mission time should be specified for a 90% reliability? (4) What should the new mean life be for a reliability of 90% and a mission time of 4 hr? Use the proper distribution (there are two distributions that will each give the same answer).

    Answers: 50; 7.7%; 5.3; 38

2. During a test of 800 parts, the number of failures in the interval between 3 and 4 hr was 46. The number of survivors at the end of the interval was 429. (1) What is the failure rate? (2) What is the mean life?

    Answers: 0.1018; 9.823

3. A certain system is normally distributed with a mean life of 90 hr and a standard deviation of 5 hr. (1) What is the chance that it

will last longer than 90 hr? (2) 100 hr? (3) 80 hr? Choose the appropriate distribution.

Answers: 50%; 2.27%; 97.7%

4. In Problem 3, (1) how many failure-free operating hours can be specified if a 95% reliability is desired? (2) 99%?

Answers: 81.8; 78.4

5. In Problem 3, (1) what new mean life must be specified for a reliability of 95% and an operating time of 80 hr? (2) What new mean life must be specified for a reliability of 99% and an operating time of 80 hr? (3) For a reliability of 99% and an operating time of 100 hr? (4) A reliability of 95% and an operating time of 100 hr?

Answers: 88.2; 91.6; 111.6; 108.2

6. A certain electronic component is normally distributed with a mean life of 2000 hr and a standard deviation of 100 hr. (1) How many of 1000 units will last longer than 1750 hr? (2) 2165 hr? (3) How many operating hours can be specified if the units must be 95% reliable? (4) What should the new mean life be for a 95% reliability and an operating time of 1750 hr?

Answers: 994; 50; 1835; 1915

7. Answer Problem 6 assuming an exponential distribution. Also solve using the binomial and compare. (The answers will be the same.)

Answers: 417; 339; 103; 34,118

8. Battery life is normally distributed with a mean life of 28 hr and a standard deviation of 3.5 hr. (1) What is the chance that a battery will last longer than 30 hr? (2) How many mission hours should be specified for a 90% reliability?

Answers: 28.4%; 23.5

9. Prove the answers to Example 6.17.

10. Prove the answers to Example 6.18.

11. Prove the answers to Example 6.19.

12. Prove the answers to Example 6.20.

13. Prove the answers to Example 6.21.

14. Solve Examples 6.2, 6.3, and 6.5 in the text using the binomial. Compare the answers to those derived using the exponential (the answers will be almost identical).

15. Solve Examples 6.18 and 6.19 using the binomial. Compare to the exponential solutions for the same examples (the answers will be almost identical).

16. Solve Problem 8 for a Weibull distribution if the Weibull slope is 3.0 and the characteristic life is 29 hr.
    Answers: 33.1%; 13.7

17. Solve Example 6.18 in the text for a mission time of 4 hr and a reliability of 99%.
    Answers:

	R (%)	T (hr)	$\overline{m}$ (hr)
1. Exponential	83.3	0.22	398
2. Normal	83.7	−19.6	45.5
3. Weibull	85.7	0.29	324

18. One hundred (100) units are tested for 100 hr each. At the end of the test, there had been 20 failures. Find (1) failure rate, (2) mean life, (3) total test hours, (4) 10-hr reliability, (5) 100-hr reliability, (6) 1000-hr reliability, (7) 100-hr probability of failure. Find each of the above values for each of the following conditions.
    (a) The items were repaired and replaced as they failed.
    (b) The failed items were not repaired and replaced, but no failure times were kept.
    (c) The items were not repaired and replaced and the total test time for the 20 failed items was 900 hours.
    Answers:

	a	b	c
1.	0.002	0.00222	0.00225
2.	500	450	445
3.	10,000	9000	8900
4.	0.980	0.978	0.978
5.	0.819	0.801	0.799
6.	0.135	0.109	0.105
7.	0.181	0.199	0.201

7

) exponential
$R = e^{-\lambda T}$

$\overline{M} = 2000 \text{ hrs} = 1/\lambda$

$\lambda = .0005$

$R = e^{-.0005(1750)}$

$= 0.417$

$R = e^{-.0005(2165)}$

$= 0.339$

$0.95 = e^{-.0005 T}$

$T = 102.6 \text{ hr}$

$.95 = e^{-\lambda \cdot 1750}$

or $\lambda = .0000293$

or $m = 34,118.$

binomial

$R = (1 - \lambda)^T$

$= (1 - .0005)^{1750} = 0.417$

$R = (1 - .0005)^{2165}$

$= .339$

$.95 = (1 - .0005)^T$

$\log .95 = T (\log .9995)$

$T = \dfrac{\log .95}{\log .9995} = 102.6$

$\overline{m} = 1/\{x\}^{1-} \exp\{(1 - R)^{1/T}\}$

CHAPTER 7 | SYSTEM RELIABILITY

The purpose of this chapter is to explain the use of static and dynamic models in predicting the reliability of complex systems. These models are also used to assist in the engineering and design of these systems.

Objectives

1. Know the product rule and how to use it to calculate the reliability of series systems.
2. Understand the concept of redundancy in parallel systems and be able to calculate the reliability of such systems.
3. Be able to calculate the reliability of series–parallel (combination) systems.
4. Understand how series and parallel reliabilities compare as systems increase in complexity.
5. Understand the difference between static and dynamic systems and be able to calculate dynamic reliability using two representative models.

7.1 Series Reliability

Series reliability is the first of the three static models presented in this chapter. A static system is defined as that system where the failure of one component has no effect on the probability of any other component failing. A series system is defined as a complex system of independent units connected together, or interrelated, in such a way that the entire system will fail if any one of the units fail. Thus, the system can be no better than its weakest component. (A chain can be no stronger than its weakest link.) Series reliability is calculated using the product rule.

The Product Rule

If a system has n components, each with a reliability P_1, P_2, ..., P_n, the reliability of the system is calculated by

$$R_s = P_1 \times P_2 \times P_3 \times \cdots \times P_n$$

where

R_s = probability of the system functioning as intended for the time intended under the conditions specified

$P_1, P_2, ..., P_n$ = probability that a component of the system will function properly for the time intended under the conditions specified

Note that, in the above rule, a time limit is not specified. Although time units are not normally given in these models, they are certainly implied. It is usually helpful, in these cases, to assume a 100-hr reliability. From now on, in this chapter, 100-hr reliability will be assumed unless otherwise specified.

Example 7.1 Three components A, B, and C have probabilities of 0.92, 0.95, and 0.96, respectively. What is the system reliability?

Solution:

$$R_s = P_1 \times P_2 \times P_3$$

$$= 0.92 \times 0.95 \times 0.96 = .839 \text{ or } 83.9\%$$

Equivalent Component Reliability

Where component reliabilities are equal to each other, the product rule equation reduces to

$$R_s = (P_c)^n$$

where

$$P_c = \text{component probability}$$

When component reliabilities are not equal, it is often helpful to establish a P_c using the geometric mean. Then, when new components are added (or replaced), they can often be assumed to have the P_c of the system. Even if the new part reliability is different, it usually will not materially affect the overall reliability of a large system; similar parts usually have similar reliabilities, and any small differences that do result will usually not change the overall reliability. Also, in any large system, the concept of compensating variations will usually affect any differences; a difference in one part will be offset by an opposite deviation in another.

Example 7.2 Using the data of Example 7.1, calculate the P_c and R_s.

Solution:

$$P_c = \sqrt[n]{P_1 \times P_2 \times \cdots \times P_n}$$

$$P_c = \sqrt[3]{0.92 \times 0.95 \times 0.96} = 0.9432$$

and

$$R_s = (P_c)^n = (0.9432)^3 = 0.839 \text{ or } 83.9\%$$

Several different types of parts are frequently grouped into families and their respective P_c's multiplied together to get the overall system reliability. In this way, replacements of similar parts can be made without materially affecting the system reliability.

Example 7.3 A system has three different types of parts A, B, and C connected together in series. Component A has 20 parts in series, B has 10 parts in series, and C has 30 parts in series. $P_A = 0.99$, $P_B = 0.98$, and $P_C = 0.999$. What is the system reliability?

Solution:

$$R_s = (P_A)^n \times (P_B)^n \times (P_C)^n$$

$$R_s = (0.99)^{20} \times (0.98)^{10} \times (0.999)^{30}$$

$$= 0.8179 \times 0.817 \times 0.970$$

$$= 0.648 \text{ or } 64.8\%$$

Note that as the number of parts increase in a series system, the system reliability decreases. In a series system, the number of parts should be kept to a minimum in order to maximize reliability.

Unreliability

Unreliability is defined as $1 - $ reliability. Thus, the unreliability of a component is

$$U = 1 - P_c \tag{7.1}$$

and the unreliability of a series system is

$$U = 1 - (P_1 \times P_2 \times P_3 \times \cdots \times P_n) \tag{7.2}$$

or

$$U = 1 - (P_c)^n \tag{7.3}$$

In Eq. (7.3), P_c can be actual component reliability where all component reliabilities are equal, or it can be calculated using the geometric mean.

Example 7.4 Find the unreliability of Example 7.1.

Solution:

$$U = 1 - (P_A \times P_B \times P_C)$$

$$= 1 - (0.92 \times 0.95 \times 0.96)$$

$$= 1 - 0.839 = 0.161 \text{ or } 16.1\%$$

Example 7.5 Find the unreliability of Example 7.3.

Solution:

$$U = 1 - (P_A^n \times P_B^n \times P_C^n)$$

$$U = 1 - (0.99^{20} \times 0.98^{10} \times 0.999^{30})$$

$$= 1 - 0.648 = 0.352 \text{ or } 35.2\%$$

Series System Reliability using λ

It is not always necessary to have, or to compute, the individual reliabilities of each component.

Suppose a series system is composed of many units, each with different failure rates. At the same time, individual component reliability is not needed. In this case, it is only necessary to sum the individual component failure rates, and then substitute the values into the basic reliability formula to compute the overall system reliability. The formula is developed as follows:

$$R_s = P_1 \times P_2 \times \cdots \times P_n$$

$$= e^{-\lambda_1 T} \times e^{-\lambda_2 T} \times \cdots \times e^{-\lambda_n T}$$

$$= e^{-(\lambda_1 T + \lambda_2 T + \cdots + \lambda_n T)}$$

$$= e^{-T(\lambda_1 + \lambda_2 + \cdots + \lambda_n)}$$

Example 7.6 A series system is composed of four components with failure rates of 0.002, 0.001, 0.0025, and 0.0005. What is the 100-hr reliability of the system?

Solution:

$$e^{-100 (0.002 + 0.001 + 0.0025 + 0.0005)} = e^{-100 (0.006)} = e^{-0.6} = 0.5488$$

Using individual reliabilities:

$$e^{-0.002 (100)} \times e^{-0.001 (100)} \times e^{-0.0025 (100)} \times e^{-0.0005 (100)}$$

$$= 0.8187 \times 0.9048 \times 0.7788 \times 0.9512 = 0.5488$$

7.2 Parallel Reliability

Parallel reliability is the second of the three static models. A parallel system is defined as a complex set of interrelated components connected in such a way that a redundant, or standby, part can take over the function of a failed part to save the system; that is, there is more than one means of accomplishing a given task. Parallel systems are often called redundant systems or standby systems. Redundancy, in reliability, refers to the use of more than one part for the same function (to take over in case the first part fails). The calculations for parallel reliability are somewhat more complex than those for series reliability and include the use of the unreliability concept. The equation is

$$R_s = 1 - (U_1 \times U_2 \times U_3 \times \cdots \times U_n) \tag{7.4}$$

or, using equivalent component unreliability,

$$R_s = 1 - (U_c)^n \tag{7.5}$$

or, using $U_c = 1 - P_c$,

$$R_s = 1 - (1 - P_c)^n \tag{7.6}$$

The reliability of a redundant system must be determined by first calculating the probability that the system or part will fail (the probability that the system or part will not last for T hours without a failure, or $1 - P_s$). Once this is determined, the reliability of the system (P_s or R_s) is just $1 - P_0$ (or $1 - U$). A system with three redundant parts (A, B, and C), for instance, will fail *if and only if* all three parts fail. Therefore, the probability that the entire system will fail is the probability that A and B and C have all failed or $P_f = P_{f_A} \times P_{f_B} \times P_{f_C}$ (or $U = U_A \times U_B \times U_C$). Now, after the probability of failure has been computed for the whole system (or, in other words, the system unreliability), the system reliability can be determined, obviously, by subtracting the unreliability from 1.

Example 7.7 Find the reliability of the system in Example 7.1 if the three components A, B, and C are now connected in parallel.

Solution:

$$R_s = 1 - (1 - P_A)(1 - P_B)(1 - P_C)$$

$$R_s = 1 - (1 - 0.92)(1 - 0.95)(1 - 0.96)$$

$$= 1 - (0.08 \times 0.05 \times 0.04)$$

$$= 1 - 0.00016 = 0.99984 = 99.98\%$$

Example 7.8 If $P_c = 0.70$, what is the reliability of the following redundant systems?

1. Two components: $\begin{array}{c} \mathrm{0.70} \\ \mathrm{0.70} \end{array}$

Solution:

$$R_s = 1 - (1 - P_c)^n$$

$$R_s = 1 - (1 - 0.70)^2 = 1 - (0.30)^2$$

$$= 1 - 0.09 = 0.91 \text{ or } 91\%$$

2. Three components: $\begin{array}{c} \mathrm{0.70} \\ \mathrm{0.70} \\ \mathrm{0.70} \end{array}$

Solution:

$$R_s = 1 - (1 - 0.70)^3 = 1 - (0.3)^3$$

$$= 1 - 0.027 = 0.973 \text{ or } 97.3\%$$

3. Four components: $\begin{array}{c} \mathrm{0.70} \\ \mathrm{0.70} \\ \mathrm{0.70} \\ \mathrm{0.70} \end{array}$

Solution:

$$R_s = 1 - (1 - 0.70)^4 = 1 - (0.3)^4$$

$$= 1 - 0.0081 = 0.9919 \text{ or } 99.2\%$$

It is evident from Example 7.8 that parallel reliability increases as the number of components increases. This is exactly opposite of series systems, where reliability decreases as the number of components increases.

Another characteristic of parallel systems, as revealed in Example 7.8, is the decrease in marginal probability. Marginal probability, in this case, refers to the increase in reliability as components are added. Reliability increased by 6.3 percentage points when a third component was added, but only increased by 1.9 percentage points upon addition of the fourth component. Obviously, a point is soon reached when the increase in the reliability is so small that it does not justify the cost of more redundant parts.

As redundancy is increased in parallel systems, it is important to balance the costs involved. The cost of adding another component must never be greater than the benefits received from the increased reliability. Although these benefits are usually equated with increased profit or decreased costs, they do not have to be limited to monetary considerations. Safety is often so paramount (aircraft and space pro-grams, for instance) as to override most cost considerations.

7.3 Series–Parallel Systems

Very complex static systems include both series and parallel components. Reliability for these systems is determined by computing the reliabilities separately, using the rules that apply to either series or parallel systems, level by level, until the entire system is completed. The steps are:

1. Separate all subsystems and categorize them as series or parallel.
2. Calculate the reliability of each parallel subsystem.
3. Calculate the reliability of each series subsystem.
4. Use each series and/or parallel subsystem as if each are units in a large, higher-level subsystem and calculate the reliability as before.
5. Continue level by level until the system is complete.

Example 7.9

1. A system has 100 parts and three branches, A, B, and C.
2. Branch A has 20 parts in series with each of its parts $P(0.95)$ or $P_A = 0.95$
3. Branch B has 20 parts in series with $P_B = 0.93$
4. Branch C has 60 parts in series with $P_C = 0.96$
5. Branches B and C are each in series with A, but parallel to each other.
6. What is the system reliability?

Solution:

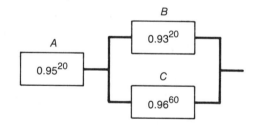

1. $R_A = (P_A)^n = 0.95^{20} = 0.358$
2. $U_B = 1 - (P_B)^n = 1 - 0.93^{20} = 0.766$
3. $U_C = 1 - (P_C)^n = 1 - 0.96^{60} = 0.914$
4. $U_{BC} = U_B U_C = (0.766)(0.914) = 0.700$
5. $R_{BC} = 1 - U_{BC} = 1 - 0.700 = 0.300$
6. $R_s = R_A R_{BC} = (0.358)(0.300) = 0.107$ or 10.7%

High Level versus Low Level

High-level redundancy refers to parallel systems, while low-level redundancy is limited to parallel components. In system redundancy, the entire system is in parallel, not the individual components. Thus, when one component fails, the entire system fails, and the entire system with all of its components (whether good or bad) must be replaced. In component (low-level) redundancy, only the individual components are parallel to each other so that, when a component fails, only the failed component needs to be replaced — it does not affect the entire system. Low-level redundancy is generally preferred over high-level redundancy because of its higher reliability and lower replacement cost.

Example 7.10

High-level redundancy:

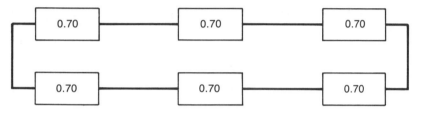

Solution:

$$R_s = 1 - \{[1 - (0.70 \times 0.70 \times 0.70)] \times [1 - (0.70 \times 0.70 \times 0.70)]\}$$

$$= 0.5684 \text{ or } 56.84\%$$

Low-level redundancy:

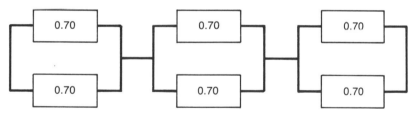

Solution:

$$R_s = [1 - (1 - 0.70)(1 - 0.70)]^3 = 0.7536 \text{ or } 75.36\%$$

7.4 Comparison of Series and Parallel Systems

As the number of components in a series system is increased, the system reliability decreases at an increasing rate. Eventually, the reliability of the system becomes so poor that the system is all but inoperable.

For any series system with more than 20 parts, the part reliabilities must be greater than 95% for the system to have much usefulness. This effect is accelerated as the system increases in complexity. As the number of parts increase, the part reliabilities must be increased in order for the system to have a reasonable level of reliability.

As the number of components is increased in parallel systems, the effect is exactly opposite to that of series systems. The system reliability increases at a decreasing rate until the point is reached where the benefits no longer offset the added costs. If part component reliabilities are rather high, only a few redundant parts are needed. But if the part reliability is rather low, many redundant parts may be needed to maximize system reliability. In any case, redundancy very quickly reaches the point of diminishing returns. Redundancy of more than 10 units, no matter what the individual probability, has little effect on the system reliability. Engineers have found that it is always best to first work with and improve part reliability before considering more redundancy.

In comparison, as numbers of components increase, series system reliability decreases while parallel system reliability increases. On the other hand, if part reliability is increased, both series and parallel systems are affected in the same way; they increase in reliability. This is one reason reliability improvement efforts should first be attempted with the parts before considering redundancy.

Example 7.11

Series systems:

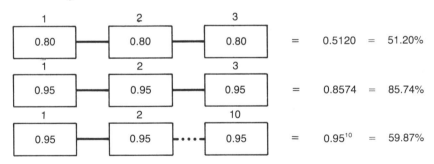

Parallel systems:

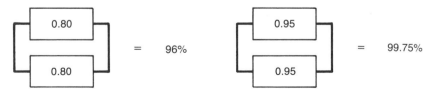

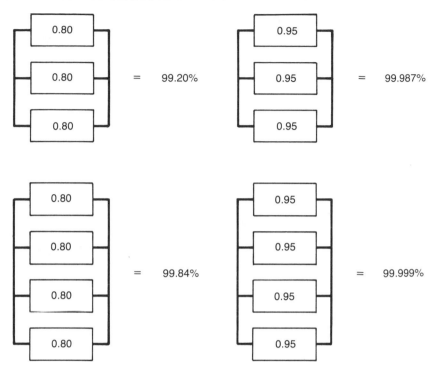

Standby Redundancy

A standby system is a parallel system where the redundant systems
or components are not energized, but are standing by waiting to be
used. When the basic (operating) system fails, the standby system is
switched into use in place of the failed system (or component). There
are two differences between a standby system and a single parallel
system. First, the standby system must have a switch which itself
must have a reliability. The reliability of the switch, if less than 100%,
will reduce the reliability of the standby system, sometimes drastically.
Second, the standby system is not energized as is the simple parallel
system and thus does not have the same opportunity of failure. Therefore, a standby system will have a higher reliability than does a simple
parallel system (assuming 100% reliability on the switch). The formula
for the standby system, is:

$$R_s = \frac{\lambda_2 \lambda_3 \cdots \lambda_n \, e^{-\lambda_1 T}}{(\lambda_2 - \lambda_1)(\lambda_3 - \lambda_1) \cdots (\lambda_n - \lambda_1)}$$

$$+ \frac{\lambda_1 \lambda_3 \cdots \lambda_n \, e^{-\lambda_2 T}}{(\lambda_1 - \lambda_2)(\lambda_3 - \lambda_2) \cdots (\lambda_n - \lambda_2)}$$

$$+ \cdots \text{(for all } \lambda\text{'s up to } \lambda_n)$$

Example 7.12 A standby system has two standby units (three units in all) with $\lambda_1 = 0.002$, $\lambda_2 = 0.001$, and $\lambda_3 = 0.003$. Assuming 100% reliability of the switch, what is the 100-hr system reliability?

Solution:

$$R_s = \frac{(0.001)(0.003)\, e^{-0.002\,(100)}}{(0.001 - 0.002)(0.003 - 0.002)}$$

$$+ \frac{(0.002)(0.003)\, e^{-0.001\,(100)}}{(0.002 - 0.001)(0.003 - 0.001)}$$

$$+ \frac{(0.001)(0.002)\, e^{-0.003\,(100)}}{(0.001 - 0.003)(0.002 - 0.003)}$$

$$= -2.4562 + 2.7145 + 0.7401 = 0.9991$$

In comparison, a simple parallel system, with the same failure rates, has a reliability of

$$1 - [(1 - e^{-0.002\,(100)}) \times (1 - e^{-0.001\,(100)}) \times (1 - e^{-0.003\,(100)})] = 0.9955$$

If the failure rates are equal ($\lambda_1 = \lambda_2$, etc.), the denominator of the above equation reduces to zero and the equation cannot be used. (The equation is automatically undefined.) In this case, another formula must be used, as follows:

$$R_s = e^{-\lambda T}[1 + \lambda T + (\lambda T)^2/2! + \cdots + (\lambda T)^{n-1}/(n-1)!]$$

Example 7.13 Calculate the 100-hr reliability of a three-component standby system where each component has a reliability of 0.002. Switching is 100% reliable.

Solution:

$$R_s = e^{-0.002(100)}\{1 + (0.002)(100) + [(0.002)(100)]^2/2\cdot 1\} = 0.9989$$

Comparing to the simple, on-line, parallel system,

$$R = 1 - [1 - e^{-0.002(100)}]^3 = 0.9940$$

Note that a simple two-component formula can be derived as a special case of the above formula:

$$R_s = e^{-\lambda T}(1 + \lambda T)$$

where

$$\lambda_1 = \lambda_2$$

7.5 Dynamic Systems

In dynamic systems, the components are dependent; the failure of one component will affect the probability of failure of other components.

Series Dynamic Systems

If dependent components are in series, the failure rate of the system is calculated by summing the reciprocals of the means (MTBF) of the components. This is the same as summing the component failure rates. Once the system failure rate is determined, the system reliability can be calculated in the normal way ($R_s = e^{-\lambda T}$).

Example 7.14 Calculate λ and R_s for the following system (100-hr reliability):

Subsystem	MTBF
1	5000
2	6000
3	4500
4	2200
5	8650

Solution:

$$\lambda = 1/m_1 + 1/m_2 + 1/m_3 + 1/m_4 + 1/m_5$$

$$= 1/5000 + 1/6000 + 1/4500 + 1/2200 + 1/8650$$

$$= 0.00116 \text{ failures per hour}$$

$$R_s = e^{-\lambda T}$$

$$R_s = e^{-0.00116\,(100)} = 0.8906 \text{ or } 89.06\%$$

Parallel Dynamic Systems

Two types of parallel dynamic systems will be illustrated: manual switching and electronic switching.

The equation for the manual switching model is

$$R_s = e^{-\lambda T}(1 + P_{sw}\lambda T)$$

Example 7.15 An electronic system has a failure rate (λ) of 0.024 failures per hour. A second, redundant system can be switched in as needed by the operator. If the operator has a 100% chance of detecting a system failure (perfect switching model), what is the 24-hr reliability for the total parallel system ($P_{sw} = 1.00$)?

Solution:

$$R_s = e^{-\lambda T}(1 + P_{sw}\lambda T)$$

$$R_s = e^{-0.024\,(24)}[1 + (1.00)(0.024)(24)]$$

$$= e^{-0.576}(1 + 0.576)$$

$$= 0.8859 = 88.59\%$$

What is the 24-hr reliability if the operator has only an 80% chance of detecting a system failure ($P_{sw} = 0.80$)?

Solution:

$$R_s = e^{-0.024(24)} [1 + 0.80 (0.024) (24)] = 0.8212 \text{ or } 82.12\%$$

When electronic switching is added to a parallel system, the equation becomes

$$R_s = e^{-\lambda T} [1 + (\lambda/\lambda_s) (1 - e^{-\lambda_s T})]$$

Example 7.16 An electronic switch has been added to the system of Example 7.15. The switch has a failure rate of 0.03 fph. What is the 24-hr reliability?

Solution:

$$R_s = e^{-0.024 \, (24)} [1 + (0.024/0.030) (1 - e^{-0.03 \, (24)})]$$

$$= e^{-0.576} [1 + (0.8) (1 - e^{-0.72})]$$

$$= e^{-0.576} [1 + (0.8) (0.5132)]$$

$$= (0.56214) (1.4106)$$

$$= 0.793 \text{ or } 79.3\%$$

Terms and Definitions

1. **Series reliability.** The probability that a system in series will function for a specified amount of time with no more than a specified number of failures. As the number of units in series are increased, the series reliability decreases.
2. **Equivalent component reliability.** When all component reliabilities are either equal or can be assumed to be equal. Unequal reliabilities are made equal by use of the geometric mean.
3. **Unreliability.** One minus the reliability. The reliability of a parallel system is one minus the product of the element (part) unreliabilities.
4. **Parallel reliability.** The probability that a parallel system will function for a specified amount of time with no more than a specified number of failures. As the number of parallel units are increased, parallel reliability will increase.
5. **Redundancy.** More than one unit to perform the same function in case the first one fails.
6. **Series system.** The elements of a system are connected together in such a way that the failure of one part (or element) causes the entire system to fail.
7. **Parallel systems.** The elements of a system are connected together

in such a way that the failure of one part is immediately compensated for by another, equal, part connected parallel to the first, failed, part.

8. **Product rule.** The computation of series reliability by successive multiplication of the individual component reliabilities: $R = P^n$.

Practice Problems

1. A system consists of 100 components connected functionally in series. Each component has a 100-hr reliability of 0.99. Calculate the reliability of the system.

   $R_s = (P_c)^n = (0.99)^{100} = 0.366$

   Answer: 0.366

2. A system consists of three components. Component 1 has 50 parts, each with a reliability of 0.98. Component 2 has 75 parts each with a reliability of 0.95, and component 3 has 100 parts with a reliability of 0.999. All the parts for each component are functionally connected in series. Calculate the system reliability if P_2 and P_3 are functionally in parallel and P_1 is in series with this combination.

   $u_2 = 1 - .95^{75} = 0.979$
   $u_3 = 1 - .999^{100} = .095$

   Answer: 33% $R_1 = (0.98)^{50} = 0.364$

   $R_{2,3} = 1 - .979 \cdot .095 = 0.907$

   $R_T = (.364)(.907) = 0.33$

3. Calculate the reliability for each of the two connections and compare the results:

   $R_1 = .95 \times .99 \times .90 \times .96 = .813$

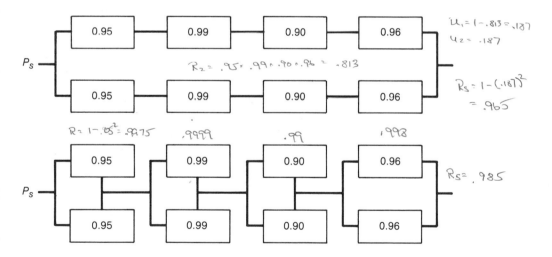

   $R_2 = .95 \times .99 \times .90 \times .96 = .813$

   $u_1 = 1 - .813 = .187$
   $u_2 = .187$

   $R_s = 1 - (.187)^2 = .965$

   $R = 1 - .05^2 = .9975$ $.9999$ $.99$ $.998$

   $R_s = .985$

   Answer: 0.9649; 0.9858

4. A radar ranging and detecting device has a failure rate of 12 failures per 1000 hr. A second system can be switched in as needed. For a manual switching device, what is the 50-hr reliability for the parallel system, assuming the operator has a 100% chance of detecting a

   $\lambda = 12/1000 = .012$ 5/h
   $\lambda_2 = 2.$ $2\bar{t} = .012(50) = 0.6$
   $R_s = e^{-\lambda t}\left[(1+\bar{t}\frac{2}{5\omega}\frac{2\bar{t}}{1})+\frac{2}{3}\right]$

   a) $R_s = e^{-.6}(1 + 1(.6)) = 0.878$
   b) $R_s = 0.549(1 + .9(.6)) = 0.845$
   c) $R_s = 0.549(1 + .5(.6)) = 0.714$

failure? A 90% chance? A 50% chance? (Assume exponential failure times.)

Answers: 87.8%; 84.5%; 71.3%

5. An electronic switch has been added to the radar unit in Problem 4 with a failure rate of 1 failure per 10,000 hr. What is the 50-hr reliability? 24 hr? $\lambda s = .0001$

Answers: 87.7%; 96.5% $R_s = 0.549(1 + \frac{.012}{.0001}(1 - e^{-.0001(50)})) = .878$

$R_s = e^{-.012(24)}(1 + \frac{.012}{.0001}(1 - e^{-.0001(24)})) = .965$

6. What is the *high*-level and *low*-level redundancy of

0.95	0.80	0.75	0.90
0.95	0.80	0.75	0.90

$R_H: R_1 .513 \quad u = .48)$
$R_2 .513$
$R_{HS} = 1 - (.487)^2 = .763$

Answers: 0.7628; 0.8887

$R_L: R_1 = 1 - .05^2 = .9975 \quad R_2 = .96 \quad R_4 = .99$
$R_3 = .9375$
$R_{SL} = .8888$

7. What is the reliability of the following complex system?

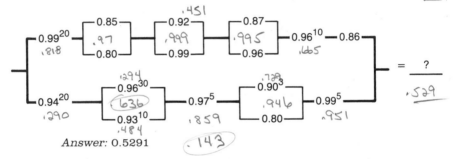

.451

0.99^{20} .818 — 0.85 / .97 / 0.80 — 0.92 / .999 / 0.99 — 0.87 / .995 / 0.96 — 0.96^{10} .665 — 0.86

0.94^{20} .290 — .294 / 0.96^{30} 636 / 0.93^{10} .484 — 0.97^5 .859 — .723 / 0.90^3 .946 / 0.80 — 0.99^5 .951

= ? .529

Answer: 0.5291   .143

8. A series system has 5 component *A*'s each with a failure rate of 0.002; 12 component *B*'s with a 0.001 failure rate each; and 20 component *C*'s with a 0.0007 failure rate each. Find the system failure rate and the 100-hr system reliability. (Assume exponential failure times.)

$R_s = e^{-T(\lambda_1 + \lambda_2 \ldots)} = e^{-100(5 \cdot .002 + 12 \cdot .001 + 20 \cdot .00)}$

Answers: 0.036; 0.0273
System $\lambda = .036$ $= .0273$

9. A certain system consists of three units (*A*, *B*, and *C*) connected together in series. The units have reliabilities of 95%, 80%, and 90%, respectively. Find the high-level and low-level redundancies.

Answers: 90%; 94.8%

H: .684

.95 — .80 — .90 / .95 — .80 — .90 $R = .900$

L: .95 / .95 — .80 / .80 — .90 / .90 $R = .948$

.9975   .960   .990

CHAPTER 8 | CONFIDENCE LIMITS

In this chapter, the procedures used for calculating confidence limits are presented. These limits are used in the control of reliability testing and to control manufacturing processes.

Objectives

1. Understand the concepts of confidence interval, confidence level, confidence limits, and level of significance; and know the four confidence level questions.
2. Know how to use the chi-square tables, and the normal curve approximation to the chi-square for large sample sizes.
3. Solve the four confidence limit questions for single-sided limits using the exponential, normal, and Weibull distributions.
4. Compare the distributions.
5. Solve the four confidence limit questions for double-sided limits.
6. Understand the concepts of sample size determination and the concept "parts per million," and the errors that can result from them.
7. Solve the four confidence limit questions using the binomial.

8.1 Confidence Intervals

Ideally, all items in a distribution should be measured and the mean determined from these measurements. However, this is seldom possible owing to the size of most distributions and the cost involved. Therefore, representative samples must be taken, and the mean of the distribution inferred from the samples. The problem with this is that there is always a possibility that the sample mean is not equal to the distribution mean. As the sample size increases, this possibility decreases, but an infinite sample size would have to be taken in order to give absolute assurance (100% probability) that the sample mean is equal to the universe mean. This impossible situation emphasizes the need for statistical sampling techniques.

The mean value derived from a sample is an example of a single-valued, or point, estimate of a universe parameter. Although this point estimate is the objective of the problem, it is really an incomplete solution. We also need to know just how accurate, or how precise, this estimate is. Confidence intervals are the practical solution to this problem. Confidence intervals are ranges of values within which we estimate the true mean to be. Confidence intervals are always associated with probability levels. An example of a probability statement using confidence intervals would be:

> We are 95% certain that the true universe mean lies between 125 pounds and 235 pounds.

An equivalent statement would be:

> If we place confidence intervals on the means of a large number of samples, the intervals of about 95% of these samples will contain the true universe mean.

The limits defining the interval are called "confidence limits" and are just the highest and the lowest values in the interval. Confidence limits of the mean can also be defined as the minimum and maximum acceptable values of the mean. The probability level associated with the confidence interval is called the "confidence level." In the preceding probability statements, the confidence level is 95% and the confidence interval is 110 lb (235 − 125). The confidence level, then, is the probability that the true mean actually lies between the confidence limits. There is always an associated probability that this is not so. This probability, the chance that the real mean does *not* lie within the calculated confidence interval, is called "the level of significance." The level of significance is 1 − confidence level. In the preceding illustration, the level of significance is 0.05 (1 − 0.95).

The preceding concepts are all illustrated in Fig. 8.1. The confidence level is the area under the curve between the lower and upper confidence limit and is equal to 1 − α. The confidence interval is the difference between the upper and lower confidence limits. The level of significance (α) is the area under the curve outside the confidence limits.

In reliability testing, the upper confidence limit is seldom needed. It is most desirable to have all reliability values (mean life, mission time, reliability, and confidence level) as high as possible — the higher, the better. Therefore, only one-sided, or one-tailed, confidence inter-

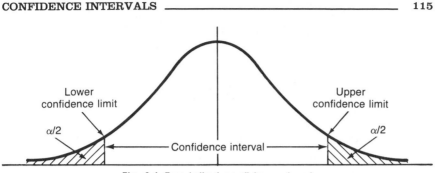

Fig. 8.1. Two-tailed confidence level.

vals are used, usually, in setting up controls for mean life. Figure 8.2 illustrates this concept. The confidence level is now the area under the curve above the lower limit, while the level of significance is the area of the lower tail below the lower limit. There may be times, however, when two-sided limits may be desired. Management may occasionally wish to set an upper design limit on engineering for cost considerations. Designs that are too good (better than they need to be) are frequently very costly.

The problem is to be reasonably certain that the true mean lies between certain limits for a given or computed mean life (or above the lower limit for one-sided limits). The only way to increase this probability is to widen the interval (or increase the upper area for one-tailed limits) or to increase the mean life (which can, and usually does, lead to increased production costs). This leads to the following four confidence limit questions:

1. What is the mean life that will be needed for a specified confidence level, reliability, and mission time?
2. What is the necessary mission time for a given confidence level and reliability? (The mean must also be used but it can be either calculated from a test sample or it can be assumed.)

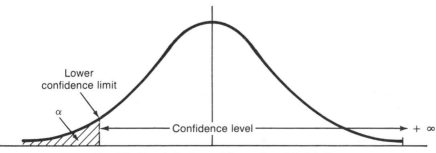

Fig. 8.2. One-tailed confidence level (left tail).

3. What is the probability (the confidence level, g) that the real universe mean lies above a specified value, or between two values, given a specified reliability (the mean life is also needed)?
4. What is the reliability limit, or limits, for a given confidence level and mission time (the mean life is also needed)?

In Chapter 6 we learned that the exponential can be used with both time-terminated and failure-terminated data. However, once the mean life has been determined, the procedures for calculating confidence limits are identical, whether the data are time-terminated or failure-terminated. Mean life, of course, can be determined in any of the ways already explained in Chapters 1 and 2. However, the mean life determined in confidence limit question 1 is a calculated value (what manufacturing must produce to in order to meet reliability and confidence level specifications) that is independent of any test or sample calculation.

The four distributions used in Chapter 6 will be used again here as will the failure-terminated data example (Example 6.6) used in Chapter 6. Thus, the distributions and procedures can be compared with Chapter 6. Since the binomial solutions are identical to the exponential, they will be given in a separate section at the end of the chapter and will not be compared in Section 8.4.

All the distributions can be used with sample data or with assumed distribution parameters (the mean only for the exponential; the mean and standard deviation for the normal; and the slope and characteristic life for the Weibull). Distribution parameters can be assumed by (1) comparing to a similar product produced by similar processes, (2) obtaining information from a long-term stable process where the product's data over a long time can be assumed to be the universe parameters, or (3) making an educated guess.

Since the chi-square is needed for calculating these confidence limit values, the appropriate chi-square value must first be determined. The chi-square (χ^2) distribution provides a means for adjusting the limits for various sample sizes. In reliability, the number of failures is used instead of the sample size. (The sample size is just used to determine the degrees of freedom and, in reliability, the degrees of freedom are determined by the number of failures.) When much sample information is available (many failures have been observed and their times to failure logged), the confidence interval can be narrow (the limits can be closer to the mean). The more information we have (the more failures we observe and measure), the more confidence we can have that our test results are accurate, that they correctly predict the real universe values.

Reliability control limits can be thought of as a form of safety margin, with the margin equaling the difference between the mean life and the lower limit (m_{LC}), and the confidence level being the probability that this margin will not be exceeded. This safety margin must be large enough to be meaningful but not so large that costs become excessive. Using the chi-square in conjunction with the applicable distribution provides a solid probability estimate for this margin rather than just a simple guess. In addition, control limit values are automatically adjusted to fit the available information — the number of observed failures. In this way, products with less failure information (less observed failures) are provided with greater safety margins.

If no test information is available (we are predicting from specifications only, for instance), a single failure must be assumed, along with the distribution parameters already mentioned. Of course, the confidence interval is quite large (actually the maximum possible) when developed from only single-failure information. By making these assumptions, the formulas can be used to extrapolate design and production information before a design is even started.

Figure 8.3 presents a picture of the reliability single-confidence-limit logic. The area under curve b above the lower limit (m_{LC}) represents the confidence level, g. The lower limit (m_{LC}) of curve b becomes the mean of curve a. The area under curve a above the lower mission time limit (T_{LC}) represents the reliability specification (R). The area under curve a above the mission time specification (T) represents the lower reliability limit (R_{LC}). Curve a represents the applicable distribution (exponential, normal, or Weibull), while curve b represents the chi-square.

8.2 The Chi-Square Probability Distribution

The chi-square distribution shows the proportion of the total number of values in a sample that are greater than a particular area under the curve (a particular probability). The formula is

$$\chi^2 = \Sigma\left(\frac{\chi - \mu}{\sigma}\right)^2 = \Sigma z^2$$

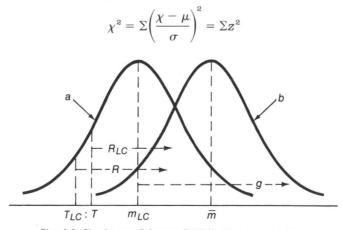

Fig. 8.3. Single-confidence-limit logic for reliability.

Table A3 is a list of the χ^2 values for degrees of freedom from 1 to 30. Degrees of freedom refer to the number of observations that are allowed to vary. (If any three numbers are to be added together to equal 9, the first two can vary, be any digit, but the third is restricted to only one number once the other two are determined. The three numbers, then, have $n - 1$ or two degrees of freedom.) In order to read the tables, only two values are needed; the degrees of freedom $2f$ and the desired probability level. For a one-tailed confidence limit, the chi-square value is found by

$$\chi^2_{2f,\alpha}$$

where

[handwritten note: If no failures have occurred, assume one up you can come with a χ^2 value]

χ^2 = chi-square

$2f$ = 2 times the number of failures in the sample

α = the predetermined level of significance (determined by engineering requirements or management decision)

The procedure is

1. Determine the number of failures (f for λT).
2. Enter the table at $2f$ under the degree of freedom column.
3. Move to the right until the α column is found.
4. The value at this point is the desired χ^2 value (it may be necessary to interpolate).

Example 8.1 Find the chi-square value for $f = 10$ and $\alpha = 0.05$

Solution:

Degrees of freedom = $2f = 2(10) = 20$; at $\alpha = 0.05$ and $df = 20$, $\chi^2 = 31.410$ (from Table A3).

Inverse Use of the Chi-Square

It is frequently desirable in reliability to find an unknown confidence limit from a given control limit and a known failure rate. The first step is to calculate the chi-square value through the inverse use of the control limit formula (see Sections 8.5 and 8.6). Once the chi-square value is known, the unknown confidence level can be determined by interpolation from the table.

Example 8.2 What is the confidence level for $f = 10$ when the chi-square value is 33.00?

Solution:

Find 33.00 in the body of the table (Table A3) and on line 20 ($2f = 20$ degrees of freedom). Since there is no 33, it will be necessary to interpolate (see Chapter 2).

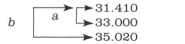
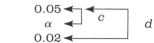

$$b \quad \begin{array}{c} a \end{array} \begin{array}{l} 31.410 \\ 33.000 \\ 35.020 \end{array} \qquad \begin{array}{l} 0.05 \\ \alpha \\ 0.02 \end{array} \begin{array}{c} c \end{array} \quad d$$

$c = da/b = [(0.05 - 0.02)(31.410 - 33.000)] \div (31.410 - 35.020)$

$\quad = [(0.03)(-1.59)] \div (-3.61) = 0.0132$

$\alpha = 0.05 - 0.0132 = 0.0368$

Confidence level $= 1 - \alpha = 1 - 0.0368 = 0.9632$ or 96.32%

Normal Curve Approximation to χ^2

When the degrees of freedom $(2f)$ are greater than 30, the table of χ^2 values cannot be used. However, as the degrees of freedom increases beyond 30, $\sqrt{2\chi^2}$ forms a normal distribution with $\sqrt{4f - 1}$ as the mean. The following normal curve approximation will apply:

$$\sqrt{2\chi^2} = \sqrt{4f - 1} + Z_{1-\alpha}$$

Z, in the above formula, is identical to the Z of Chapter 2 and is equal to the number of standard deviations from the mean. Z is determined from the table of standard normal deviates (Table A1) and is based on the confidence level determined by management. Solving for χ^2 in the preceding formula gives the following, more easily used, formula:

$$\chi^2_{2f,\alpha} = \tfrac{1}{2}(\sqrt{4f - 1} + Z_{1-\alpha})^2$$

When solving inverse problems, the above formula must be solved for Z:

$$Z_{1-\alpha} = \sqrt{2\chi^2} - \sqrt{4f - 1}$$

Example 8.3 Find the chi-square value for $f = 10$ and $\alpha = 0.05$ using the normal curve approximation.

Solution:

$$Z_{1-\alpha} = Z_{1-0.05} = Z_{0.95} = 1.645 \text{ (Table A1)}$$

$$\chi^2_{2f,\alpha} = \tfrac{1}{2}(\sqrt{4f - 1} + Z_{1-\alpha})^2$$

$$\chi^2_{2f,\alpha} = \chi^2_{20,0.05} = \tfrac{1}{2}(\sqrt{4(10) - 1} + 1.645)^2 = 31.13$$

Comparing the two answers (31.41 for Example 8.1 and 31.13) shows that the two methods used give answers reasonably close to each other, at least for large sample sizes.

Inverse Use of the Normal Curve Approximation

As explained in Section 8.2, it is frequently useful to be able to calculate an unknown confidence limit from a known failure rate and a given control limit. When the degrees of freedom exceed 30, the inverse use of the normal curve approximation must be used instead of the inverse use of the chi-square table.

Example 8.4 What is the confidence level for $f = 10$ when the chi-square is 33?

Solution:

$$Z_{1-\alpha} = \sqrt{2\chi^2} - \sqrt{4f - 1}$$

$$Z_{1-\alpha} = \sqrt{2(33)} - \sqrt{4(10) - 1}$$

$$= \sqrt{66} - \sqrt{39}$$

$$= 1.879$$

$$P_z = 0.9698 \text{ or } 96.98\% \text{ (Table A1)}$$

Note: Compare this answer (96.98%) to the answer found by using the table (96.32%).

Two-Sided Chi-Square Values

Example 8.5 Find the two-sided chi-square values for a 95% confidence level where there had been 10 failures in the test. Use the normal curve approximation formula.

Solution:

note: $\alpha/2$

$$\chi^2_{2f,\alpha/2} = \chi^2_{20,0.025} = 34.42 \text{ (from Table A3)}$$

$$\chi^2_{2f,1-\alpha/2} = \chi^2_{20,0.975} = 9.341 \text{ (from Table A3)}$$

If the sample size is large (the degrees of freedom are greater than 30), the normal curve approximation must be used. For two-sided chi-square values, the formulas are

$$\chi^2_{2f,\alpha/2} = \frac{1}{2}(\sqrt{4f - 1} + Z_{\alpha/2})^2$$

$$\chi^2_{2f,1-\alpha/2} = \frac{1}{2}(\sqrt{4f - 1} - Z_{1-\alpha/2})^2$$

In Example 8.4 the chi-square values, using the normal curve approximation, are

$$\chi^2_{20,0.025} = \frac{1}{2}(\sqrt{4(10) - 1} + 1.96)^2 = 33.66$$

$$\chi^2_{20,0.975} = \frac{1}{2}(\sqrt{4(10) - 1} - 1.96)^2 = 9.18$$

8.3 Single-Sided Lower Confidence Limits

Although double-sided limits are almost never necessary for reliability, they can be calculated. This will be the subject of Section 8.5.

A set of two formulas is used to solve the four confidence limit questions. Of course, the formulas are different for each distribution.

Sometimes the formulas are used directly and sometimes inversely. The nature of the problem determines the method.

In each set of formulas, a "mean life" confidence limit is required (m_{LC}). This limit is used only as an intermediary calculation to get to the desired value. In actuality, the "mean life" limit is seldom used as a control measure and is, therefore, seldom required in the specifications. (The mission time specifications, for instance, are much more meaningful.)

One-Sided Limits using the Exponential

Exponential confidence limits must use the chi-square as well as the basic exponential formulas. The basic formulas are

$f = \# failures.$

Time truncated tests, *

$$m_{LC} = 2f\overline{m}/\chi^2_{2f,\alpha}$$

$$R_{LC} = \exp(-T/m_{LC})$$

These formulas have been solved inversely, where necessary, for the following two-step solution procedures:

1. Find the mean life, $\overline{m}$:
$$m_{LC} = T/\ln(1/R_{LC})$$
$$\overline{m} = \chi^2_{2f,\alpha}(m_{LC})/2f$$

2 sided

$m_{LC} = \dfrac{2f\overline{m}}{\chi^2_{2f,d/2}}$ note d/2

2. Find the mission time limit, T_{LC}:
$$m_{LC} = 2f\overline{m}/\chi^2_{2f,\alpha}$$
$$T_{LC} = m_{LC}[\ln(1/R_{LC})]$$

$m_{uc} = \dfrac{2f\overline{m}}{\chi^2_{2f,1-d/2}}$

3. Find the confidence level, g:
$$m_{LC} = T/\ln(1/R_{LC})$$
$$\chi^2_{2f,\alpha} = 2f\overline{m}/m_{LC}$$
$$g = 1 - \alpha$$

* for failure truncated (all samples fail)

4. Find the lower reliability limit, R_{LC}:
$$m_{LC} = 2f\overline{m}/\chi^2_{2f,\alpha}$$
$$R_{LC} = \exp(-T/m_{LC})$$

$\chi^2_{2f+2,d}$ use $2f+2$

Example 8.6 Solve Example 6.6 for the four confidence limit questions. The sample size is 6, the number of failures is 6, and the mean is 76.8. (1) Find the mean life for a 90% reliability, a 95% level of confidence and a 40 hour mission time. (2) Find the mission time for a 90% reliability, a 95% level of confidence, and the above mean life, 76.8 hr. (3) Find the confidence level for a 90% reliability, a mission time of 40 hr, and a mean life of 76.8 hr. (4) Find the reliability for a confidence level of 95%, a mission time of 40 hr, and a mean life of 76.8 hr. Note that the 90% reliability is a specified lower confidence limit, a lower "reliability" limit. When problems, or specifications, are stated this way (what is the lower limit for a reliability of 90% at a confidence level of 95%, for instance), the stated reliability percentage is always a lower limit.

Solution:

(1) The mean life, $\overline{m}$, is

$m_{LC} = T/\ln(1/R_{LC}) = 40/\ln(1/0.90) = 380$ hr

$\chi^2_{2(6),0.05} = 21.026$ (Table A3)

$\overline{m} = \chi^2_{2(6),0.05}(m_{LC})/2f = 21.026(380)/2(6) = 666$ hr

In order to be 95% confident that no more than 10% of the units will fail before 40 hr, the units must be produced so that they have an average mean life of 666 hr (if the distribution is exponential and there have been six failures).

(2) The mission time limit, T_{LC}, is

$\chi^2_{2(6),0.05} = 21.026$ (Table A3)

$m_{LC} = 2f\overline{m}/\chi^2_{2(6),0.05} = 2(6)(76.8)/21.026 = 43.8$ hr

$T_{LC} = m_{LC}[\ln(1/R_{LC})] = 43.8[\ln(1/0.90)] = 4.615$ hr

If the product is produced to an average mean life of 76.8 hr, the specified mission time must be 4.615 hr in order to be 95% confident that no more than 10% of the units will fail before the specified mission time (if the distribution is exponential and there were six failures in the test).

(3) The confidence level, g, is

$m_{LC} = T/\ln(1/R_{LC}) = 40/\ln(1/0.90) = 380$ hr

$\chi^2_{2(6),\alpha} = 2f\overline{m}/m_{LC} = 2(6)(76.8)/380 = 2.425$

$\alpha = 0.99+$ (Table A3)

$g = 1 - 0.99+ = 0.0\%$

If the product is produced to an average mean life of 76.8 hr, we can have no confidence at all that no more than 10% of the items will fail before 40 hr (assuming exponentiality and $f = 6$).

(4) The lower reliability limit, R_{LC}, is

$\chi^2_{2(6),0.05} = 21.026$ (Table A3)

$m_{LC} = 2f\overline{m}/\chi^2_{2(6),0.05} = 2(6)(76.8)/21.026 = 43.8$ hr

$R_{LC} = \exp(-T/m_{LC}) = \exp(-40/43.8) = 40.1\%$

If the product is produced to an average mean life of 76.8 hr, we can be 95% confident that no less than 40.1% of the items will survive beyond 40 hr (assuming exponentiality and $f = 6$).

Example 8.7 Solve Example 8.6 for the mean life assuming that no sample information is available ($f = 1$). This example illustrates the "predictive" capability of reliability procedures. Probable reliability values can be calculated, using this type of example, before the design is even started. Thus, valuable information is available to guide engineering in their design activity. If these probable, or predictive, values are such that extensive redesign (of the product or the process) is necessary, management may even wish to reconsider the contract (obviously, these calculations can even occur prior to signing the contract).

Solution:

The mean life, $\overline{m}$, is

$$m_{LC} = T/\ln(1/R_{LC}) = 40/\ln(1/0.90) = 380 \text{ hr}$$

$$\chi^2_{2(1),0.05} = 5.991 \text{ (Table A3)}$$

$$\overline{m} = \chi^2_{2(1),0.05}(m_{LC})/2f = 5.991(380)/2(1) = 1138 \text{ hr}$$

Note how the mean life has increased over the previous example (1138 hr instead of 666 hr). When sample information is not available, product capability must be improved dramatically to offset the increased uncertainty. This is almost certain to increase production costs. This also illustrates the importance of obtaining actual test information as quickly as possible, especially prior to manufacture (who may not need to provide the expensive processes indicated by the preliminary analysis).

Example 8.8 A certain product has a design criteria of 26 hr of operation without a failure and the estimated mean life of the product is 75 hr. (1) Find the reliability if the 26 hr are assumed to be mission hours and the distribution is assumed to be exponential. (2) Find the confidence level (g) if the 26 hr are assumed to be the lower confidence limit and the distribution is assumed to be exponential. Of course, the distribution could also be normal, but a standard deviation would also have to be assumed.

Solution:

1. $R = \exp(-26/75) = 70.7\%$
2. $m_{LC} = 26 = 2(1)(75)/\chi^2_{2(1),\alpha}$
    $\chi^2_{2(1),\alpha} = 2(1)(75)/26 = 5.769$
    $\alpha = 0.058$
    $g = 1 - 0.058 = 94.2\%$

In problems of this nature it is very tempting to use the 94.2% figure as the reliability and to so state in the proposals. However, this is incorrect. The correct procedure is to first clarify the specification (is it mission time or lower confidence limit?) and to then use the correct procedure of the two. If the specification does indeed refer to the lower confidence limit, the 94.2% should never be presented as an expected reliability goal. Confidence level does not mean reliability. It simply means that there is a 94.2% chance that the real mean life is greater than, or equal to, 26 hr. The mission time must be known in order to calculate reliability (refer to part 4 of Example 8.6).

One-Sided Limits using the Normal

The formulas for solving the four confidence limit problems using the normal are based on the two following formulas. αa refers to the area of curve a, while αb refers to the area of curve b.

$$z_{\alpha a} = (T - m_{LC})/\sigma \text{ (if sample data are used, } \sigma = s)$$

$$m_{LC} = 2f\overline{m}/\chi^2_{2f,\alpha b}$$

These formulas have been solved inversely, where applicable, for the following two-step solution procedures.

1. Find the mean life, $\overline{m}$:
   $$m_{LC} = T - \sigma(z_{\alpha a})$$
   $$\overline{m} = m_{LC}(\chi^2_{f,\alpha b})/2f$$
2. Find the mission time limit, T_{LC}:
   $$m_{LC} = 2f\overline{m}/\chi^2_{2f,\alpha b}$$
   $$T_{LC} = m_{LC} + \sigma(z_{\alpha a})$$
3. Find the confidence level, g:
   $$m_{LC} = T - \sigma(z_{\alpha a})$$
   $$\chi^2_{2f,\alpha b} = 2f\overline{m}/m_{LC}$$
   $$g = 1 - \alpha b \text{ (Table A3)}$$
4. Find the reliability lower limit, R_{LC}:
   $$m_{LC} = 2f\overline{m}/\chi^2_{2f,\alpha b}$$
   $$z_{\alpha a} = (T - m_{LC})/\sigma$$
   $$R_{LC} = 1 - \alpha a \text{ (Table A1)}$$

Example 8.9 Solve Example 8.6 using the normal. Since $\alpha a = 1 - 0.90$, $z_{\alpha a} = z_{0.10} = -1.282$ (Table A1). Since $\alpha b = 1 - 0.95$, $\chi^2_{2f,\alpha b} = \chi^2_{2(6),0.05} = 21.026$ (Table A3).

Solution:

(1). The mean life, $\overline{m}$, is
$$m_{LC} = T - s(z_{\alpha a}) = 40 - 14.6(-1.282) = 58.72$$
$$\overline{m} = m_{LC}(\chi^2_{2f,\alpha b})/2f = 58.72(21.026)/12 = 103 \text{ hpf}$$
(2). The mission time limit, T_{LC}, is
$$m_{LC} = 2f\overline{m}/\chi^2_{2f,\alpha b} = 2(6)(76.8)/21.026 = 43.83$$
$$T_{LC} = m_{LC} + s(z_{\alpha a}) = 43.83 + 14.6(-1.282) = 25.1 \text{ hr}$$
(3). The confidence limit, g, is
$$m_{LC} = T - s(z_{\alpha a}) = 40 - 14.6(-1.282) = 58.72$$
$$\chi^2_{2f,\alpha b} = 2f\overline{m}/m_{LC} = 2(6)(76.8)/58.72 = 15.695$$
$$\alpha b = 0.209 \text{ (Table A3)}$$
$$g = 1 - \alpha b = 1 - 0.209 = 79.1\%$$
(4). The reliability limit, R_{LC}, is
$$m_{LC} = 2f\overline{m}/\chi^2_{2f,\alpha b} = 2(6)(76.8)/21.026 = 43.83$$
$$z_{\alpha a} = (T - m_{LC})/\sigma = (40 - 43.83)/14.6 = -0.262$$
$$P(z_{\alpha a}) = \alpha a = 0.397 \text{ (Table A1)}$$
$$R_{LC} = 1 - \alpha a = 1 - 0.397 = 60.3\%$$

Unknown Sample Size. As in the exponential, the normal can also be used to predict the mean life when no sample information is available. A single failure ($f = 1$) and a standard deviation must first be assumed. The formulas and calculations are identical except that the chi-square is entered at two degrees of freedom instead of 12. Assuming the same standard deviation (14.6 hr), the calculations are

$$m_{LC} = T - \sigma(z_{\alpha a}) = 40 - 14.6(-1.282) = 58.72$$

$$\overline{m} = m_{LC}(\chi^2_{2f,\alpha a})/2f = 58.72(5.991)/2(1) = 176 \text{ hpf}$$

Note how the necessary mean life has increased (103 to 176) when no sample data are used. When sample information is not available, product capability must be improved dramatically to offset the increased uncertainty. This is almost certain to increase production costs. This also illustrates the importance of obtaining actual test information as quickly as possible, especially prior to manufacture (who may not need to provide the expensive processes indicated by the preliminary analysis).

One-Sided Limits using the Weibull Graph

The Weibull graph can be used to find the lower reliability limit and mission time only. It is not possible to answer the other confidence limit questions with the Weibull graph.

Example 8.10 Using the Weibull graph, find the lower reliability limit and the mission time to Example 8.6.

Solution:

Failure Times	95% Rank (Table A5)
53	39
66	58
80	73
82	85
89	94
91	99

From the graph (Fig. 8.4):

1. $R_{LC} = 78\%$
2. $T_{LC} = 23 \text{ hr}$

Graphical-confidence-limit solutions require the fitting of curves by sight to the point pairs and are therefore subject to gross errors. Also, the tables for using the graph are very extensive; Table A5 presents only a small portion of the overall tables available. The Weibull formulas, presented next, are recommended instead.

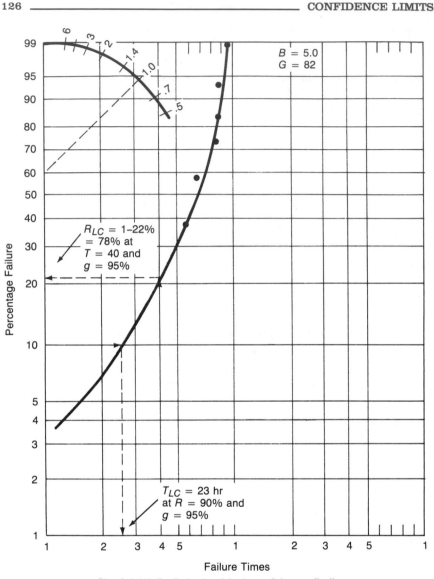

Fig. 8.4. Weibull single-sided confidence limits.

Single-Sided Limits using the Weibull Formulas

When Weibull formulas are used to determine the answers to the four confidence limit questions, the chi-square formula is included (in a two-step procedure). Where necessary, inverse solutions to the basic formulas are used. Chi-square values are found in Table A3. The two Weibull parameters, B and G, must first be determined and then the answers to the four confidence limit questions can be found.

The following formulas are empirically derived (by trial and error, mostly) rather than determined from rigorous statistical and mathematical logic. The reason for this is that the rigorous formulas are much too complex for easy use (for one thing, they involve the use of an enormous amount of tables). Although the simplified formulas given provide a good approximation of the more rigorous formulas, it must be emphasized that they are only approximations. In most cases, the approximation will be close enough for all practical purposes. In the few cases where closer precision is needed, the more theoretical formulas should be used (see your friendly statistician). One of the more compelling advantages of the simplified formulas is that they are easily programmable for use on small personal computers.

A. Find B and G. X is the time to failure and Y is the median rank value from Table A5 (or Y is derived from one of the Y formulas of Section 6.6).

$$B = \frac{[\Sigma(\ln X)\{\ln \ln[1/(1 - Y)]\}] - [(\Sigma \ln X)\{\Sigma \ln \ln[1/(1 - Y)]\}]/n}{\Sigma(\ln X)^2 - (\Sigma \ln X)^2/n}$$

$$a = \{\Sigma \ln \ln[1/(1 - Y)]\}/n - B[\Sigma \ln X]/n$$

$$G = \exp[(-0.0003277 - a)/B]$$

B. Find the four confidence limit values.

  1. Find the new mean life, $\overline{m}$:
     $$G_{LC} = \exp\{\ln T - [\ln \ln(1/R)]/B\}$$
     $$\overline{m} = (\chi^2_{2f,\alpha})G_{LC}/2f$$
  2. Find the mission time limit, T_{LC}:
     $$G_{LC} = 2fG/\chi^2_{2f,\alpha}$$
     $$T_{LC} = G_{LC}(-\ln R)^{1/B}$$
  3. Find the confidence level, g:
     $$G_{LC} = \exp\{\ln T - [\ln \ln(1/R)]/B\}$$
     $$\chi^2_{2f,\alpha} = 2fG/G_{LC}$$
     $$g = 1 - \alpha \text{ (Table A3)}$$
  4. Find the reliability limit, R_{LC}:
     $$G_{LC} = 2fG/\chi^2_{2f,\alpha}$$
     $$R_{LC} = \exp[-(T/G_{LC})^B]$$

Example 8.11 Solve Example 8.6 for the four confidence limit questions. Inputs are $R = 90\%$, $g = 95\%$, and $T = 40$ hr. $\chi^2_{2(6),0.05} = 21.026$ (Table A3). The Weibull parameters, from Example 6.16, are $B = 4.987$ and $G = 83.51$.

Solution:

  1. The new mean life, $\overline{m}$, is
     $$G_{LC} = \exp\{\ln T - [\ln \ln(1/R)]/B\}$$
     $$G_{LC} = \exp\{\ln 40 - [\ln \ln(1/0.90)]/4.987\} = 62.81 \text{ hr}$$
     $$\overline{m} = (\chi^2_{2(6),0.05})G_{LC}/2f = (21.026)62.81/2(6) = 110 \text{ hr}$$

2. The mission time limit, T_{LC}, is

$$G_{LC} = 2fG/\chi^2_{2(6),0.05} = 2(6)(83.51)/21.026 = 47.7 \text{ hr}$$

$$T_{LC} = G_{LC}(-\ln R)^{1/B} = 47.7(-\ln 0.90)^{1/4.987} = 30.3 \text{ hr}$$

3. The confidence level, g, is

$$G_{LC} = \exp\{\ln T - [\ln \ln(1/R)]/B\}$$

$$G_{LC} = \exp\{\ln 40 - [\ln \ln(1/0.90)]/4.987\} = 62.81 \text{ hr}$$

$$\chi^2_{2(6),\alpha} = 2fG/G_{LC} = 2(6)(83.51)/62.81 = 15.9545$$

$$\alpha = 0.195 \text{ (Table A3)}$$

$$g = 1 - \alpha = 1 - 0.195 = 80.5\%$$

4. The reliability limit, R_{LC}, is

$$G_{LC} = 2fG/\chi^2_{2(6),0.06} = 2(6)(83.51)/21.026 = 47.7 \text{ hr}$$

$$R_{LC} = \exp[-(T/G_{LC})^B] = \exp[-(40/47.7)^{4.987}] = 66.0\%$$

Unknown Sample Size. As in the exponential and normal, the Weibull can also be used to predict the mean life when no sample information is available. A single failure ($f = 1$) and a Weibull slope (B) must first be assumed. The formulas and calculations are identical except that the chi-square is entered at two degrees of freedom instead of 12. Assuming the same slope ($B = 4.987$), the calculations are

$$G_{LC} = \exp\{\ln T - [\ln \ln(1/R)]/B\}$$

$$G_{LC} = \exp\{\ln 40 - [\ln \ln(1/0.90)]/4.987\} = 62.81$$

$$\overline{m} = G_{LC}(\chi^2_{2f,\alpha})/2f = 62.81(5.991)/2(1) = 188 \text{ hpf}$$

Note how the necessary mean life has increased (110 to 188) when no sample data are used. When sample information is not available, product capability must be improved dramatically to offset the increased uncertainty. This is almost certain to increase production costs. This also illustrates the importance of obtaining actual test information as quickly as possible, especially prior to manufacture (who may not need to provide the expensive processes indicated by the preliminary analysis).

Simple Mean Life Limits

Reliability limits are sometimes presented as simple mean life limits without the reliability values R or T or both included, as follows.

Example 8.12 Using the data of Example 8.6, determine (1) the mean life lower limit for a confidence level of 95%; (2) the new mean life for a confidence level of 95% and a specified lower mean life confidence limit of 40 hr (the 40 hr is *not* a mission time); (3) the confidence level for a lower mean life limit of 40 hr; the mean is 76.8 and the standard deviation is 14.6 hr. Use the normal for comparison.

Solution:

1. The mean life lower limit, m_{LC}, is

$$m_{LC} = \overline{m} - \sigma Z_\alpha = 76.8 + 14.6(-1.645) = 52.8 \text{ hr}$$

We are 95% confident that the product will last longer than 52.8 hr. Although this may have some value as a production control measure, it is NOT a mission time because it does NOT include a reliability. In Example 8.9, for instance, the mission time lower limit was 34.1 hr. It was lower because it included a reliability specification.

2. The new mean life, $\overline{m}$, is

$$\overline{m} = m_{LC} - \sigma Z_\alpha = 40 - 14.6(-1.645) = 64.0 \text{ hr}$$

The product must be produced to a mean life of at least 64 hr in order to be 95% confident that the true mean life is greater than 40 hr. Actually, this value would be very misleading to production, since it does not include a reliability specification. In Example 8.9, which does include a reliability specification, the mean life is 82.7 hr, a considerably higher figure.

3. The confidence level, g, is

$$Z_\alpha = (m_{LC} - \overline{m})/\sigma = (40 - 76.8)/14.6 = -2.52$$
$$P(Z_\alpha) = 0.0059$$
$$g = 1 - P(Z_\alpha) = 1 - 0.0059 = 99.41\%$$

We are 99.41% confident that this product mean life is greater than 40 hr. If management wishes to be 95% confident, or greater, this calculation would suggest that all was well. However, once again, a reliability was not included. In Example 8.9, where a reliability specification was included, the confidence level was only 89.25%. Using the above 99.41% figure could lull management into a false sense of security.

Although the limits in Example 8.12 correctly represent the theory and procedure of confidence limits, they are not reliability limits. True reliability limits must include the reliability specifications.

8.4 Comparing the Distributions for Single Limits

The answers to Example 8.6 for the four distributions are given in Table 8.1 for easy comparison. Also compare to Section 6.7. The two extra examples from Section 6.7 will also be given in this section but the proof of the answers will be left to the reader as an exercise.

Example 8.13 A sample of 10 units was tested to failure. The failure times, in hours, were 2, 4, 8, 11, 15, 20, 25, 34, 45, and 55. The distribution is close to exponential. Solve the four confidence limit

Table 8.1 Comparison of Confidence Limits

Distribution	$\overline{m}$ (hr) ($f = 6$)	T (hr)	g (%)	R_{LC} (%)	$\overline{m}$ (hr) ($f = 1$)
1. Exponential	666	4.62	0.00	40.1	1138
2. Normal	103	25.1	79.1	60.3	176
3. Weibull	110	30.3	80.5	66.0	188

questions for the four distributions and compare the answers: (1) Find the mean life for a reliability of 90%, a confidence level of 90%, and a mission time of 2 hr. (2) Find the mission time for a reliability of 90%, a confidence level of 90%, and the preceding mean life. (3) Find the confidence level for a reliability of 90%, a mission time of 2 hr, and the preceding mean life. (4) Find the lower reliability limit for a mission time of 2 hr, a confidence level of 90%, and the preceding mean life. Compare to Example 6.18.

Distribution	$\overline{m}$ (hr)	T (hr)	g (%)	R_{LC} (%)
1. Exponential	27.0	1.6	68.7	87.8
2. Normal	35.3	−7.45	37.4	77.4
3. Weibull	24.8	1.9	87.7	89.7

Example 8.14 A sample of 20 units were tested to failure. The failure times, in cycles per use, were 450, 510, 555, 585, 610, 625, 655, 675, 690, 710, 730, 750, 765, 785, 800, 825, 850, 885, 915, and 1000. The distribution is close to normal. Solve the four confidence limit questions for the four distributions and compare the answers: (1) Find the mean life for a reliability of 95%, a confidence level of 95%, and a mission time of 500 cycles. (2) Find the mission time for a reliability of 95%, a confidence level of 95%, and the preceding mean life. (3) Find the confidence level for a reliability of 95%, a mission time of 500 cycles, and the preceding mean life. (4) Find the lower reliability limit for a mission time of 500 cycles, a confidence level of 95%, and the preceding mean life. Compare to Example 6.19.

Distribution	$\overline{m}$ (hr)	T (cycles)	g (%)	R_{LC} (%)
1. Exponential	14,223	25.3	0.00	36.2
2. Normal	1,014	287	56.5	55.2
3. Weibull	1,208	320	41.7	49.4

Example 8.15 Solve Example 8.6 for a desired confidence level of 99%, a mission time of 40 hr, and a reliability of 99%. Compare to Examples 8.6 and 6.20.

Distribution	$\bar{m}$ (hr)	T (hr)	g (%)	R_{LC} (%)
1. Exponential	8,695	0.35	0.00	32.0
2. Normal	162	1.11	54.8	37.1
3. Weibull	220	15.19	36.1	28.6

Example 8.16 Solve Example 8.6 for a desired confidence level of 90%, a mission time of 50 hr, and a reliability of 90%. Compare to Examples 8.6 and 6.21.

Distribution	$\bar{m}$ (hr)	T (hr)	g (%)	R_{LC} (%)
1. Exponential	735	5.23	0.00	36.5
2. Normal	91	30.9	79.0	74.5
3. Weibull	122	34.3	57.2	50.3

Analysis of Comparisons

The following comparisons and observations can be made from the above examples.

1. The comparisons made for Chapter 6 are equally appropriate here except that all differences among the distributions are now magnified. The mean values dealt with in Chapter 6 represent 50% levels of confidence. It should be obvious, then, that differences in the answers of the various distributions increase as confidence levels are increased.

2. If the distribution is exponential, the allowed mission time must be much lower than if the distributions were normal (all else equal). Remember, however, that only one distribution in each problem gives the correct answer. The other distributions are shown only as comparisons to emphasize the degree of error inherent in the use of the wrong distribution.

3. The correlation between the Weibull and other distributions, noted in Chapter 6, still exist, although it is not as close. This will always be the case with confidence limits. The higher the confidence level, the greater this discrepancy is likely to be.

4. As confidence levels are increased, the reliability specifications (R and T) are very strongly affected. This usually means that the product must be produced to a much higher quality, since R and T can seldom be reduced (owing to customer requirements). Obviously, increasing the confidence level can enormously increase production problems and costs.

5. It is extremely important that the proper distribution be used, especially when higher levels of confidence are specified. There are many examples in the previous comparisons, some already men-

tioned, that illustrate this necessity, but one is particularly obvious. In Table 8.1, the calculated level of confidence, g, for the exponential was 0.0%. If this distribution were in fact exponential, the product quality would have to be drastically increased in order to meet minimum requirements (which is usually a very costly project). However, this distribution is NOT exponential; it is actually closer to the Weibull. Since the Weibull shows a level of confidence of 80.5% and a calculated reliability of 66%, the product is much closer to being acceptable than the exponential answers would seem to indicate. The actual changes required, therefore, to meet the specifications will not be nearly as great as indicated by the exponential (and hopefully not nearly as costly). Obviously, the use of an incorrect distribution could lead to costly errors.

6. As in Chapter 6 (Section 6.7 "Problem Interpretation"), problems in this chapter are answered for all four confidence limit questions at the same time. In actual practice, of course, only one question will usually be able to be answered at a time because only three pieces of data input will usually be available. The fourth piece of data must be assumed in order to do the type of exhaustive management analysis described in Chapter 6 (Section 6.7). Also, as in Chapter 6, it is best if the assumed data are realistic so that the answers derived from them are usable in management decision making.

8.5 Double-Sided Confidence Limits

The formulas used for double-sided limits are very similar to those used in single-sided limits. The only difference is the way the confidence level is used. The confidence level now refers to the area under the curve between the two limits (see Fig. 8.1) not the area above the lower limit, as it did in single limits. Therefore, the level of significance, α, is now divided between the two tails, with half ($\alpha/2$) below the lower limit and half above the upper limit. Wherever the confidence level is used in double-limit formulas, then, two values can be generated, in most cases, instead of the one determined in single-limit formulas. Since the calculations must pertain to areas below the designated or critical values (the limits), these two areas are designated as $\alpha/2$ for the area below the lower limit and $1 - \alpha/2$ for the area below the upper limit (only the areas below the critical values are contained in the tables).

Two values, an upper value associated with $1 - \alpha/2$ and a lower value associated with $\alpha/2$, can be calculated for three of the confidence limit questions (1, 2, and 4). Thus, a range of actions, or a range of possibilities, is identified. The third, g, must always be a single value, since it represents the single area under the curve between the two limits.

The two values generated for mission time (confidence limit question 2) and for reliability (confidence limit question 4) derive from a single mean life value (which, of course, is either assumed or determined by testing). However, the two possible calculated mean life values (confidence limit question 1) are associated with, and would generate, a pair of mission times and reliability limits for each. Therefore, only one of the possible mean life values can be used (only one set of mission times and reliability limits can be specified). The question arises as to which. The mean life always implies an "or more than" relationship. We wish to design, and produce, to "no less than" the mean life specification. A specification that requires a product to average "no less than 100 and no less than 500 hr," for instance, makes no sense.

It should be obvious from the preceding logic that the greater of the two possible mean life values should be retained. Since this value, the greater of the two, is always derived from, and associated with, the lower limit only, only the lower limit calculation for the mean life is used. This calculation procedure has already been described in Section 8.3 for single lower limits. However, the single mean life generated for double limits will always differ from the single mean life determined for single limits (all other things equal) owing to the division of the alpha value; only half of the alpha area ($\alpha/2$) now appears below the lower limit.

Perhaps a good way to think about the mission time limits (or mean life limits as they are usually called) is as follows. The lower mean life limit may be equated, roughly, with the warranty period. We would like to limit the failures during this time to minimize warranty costs. The upper limit could be considered a desirable wear-out limit (something like planned obsolescence). The product is thus deliberately designed to fail by then so that more units can be sold.

In the following formulas, m_{LC} and m_{UC} are calculated for use as intermediary values to get to the desired values (m_{LC} is always associated with a lower limit and m_{UC} with the upper limit). The symbol R will apply to the specified reliability, while the calculated reliability limits will be designated as R_{LC} and R_{UC} (for lower and upper reliability limits, respectively). The same will be true for the mission time. The symbol for the designated, specified, mission time will be T, while the two calculated mission limits will be T_{LC} and T_{UC} (for lower and upper mission time limits, respectively).

Figure 8.5 presents a picture of the logic used. The area under the curve of distribution a between the two limits (m_{LC} and m_{UC}) represents the confidence level (g). The lower mean life limit (m_{LC}) of distribution a becomes the mean of distribution b, while the upper mean life limit (m_{UC}) becomes the mean of distribution c. The area under the curve of distribution b between T_{LC} and T_{UC} is the specified reliability

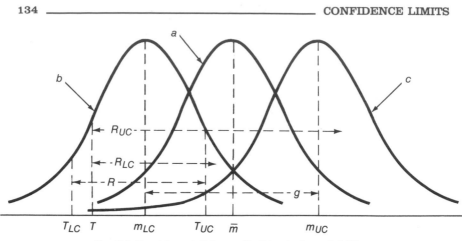

Fig. 8.5. Double-confidence-limit logic for reliability.

(R). The area under curve b above the specified mission time limit (T) represents the lower reliability limit (R_{LC}). The upper reliability limit (R_{UC}) is the area under curve c above the specified mission time limit (T). Curve b refers to the chi-square, while a and c refer to the applicable distribution (normal, exponential, or Weibull).

Two-Sided Limits using the Exponential

Exponential confidence limit formulas are based on the following six basic formulas:

1. $m_{LC} = T/\ln(1/R)$
2. $m_{LC} = 2f\overline{m}/\chi^2_{2f,\alpha/2}$
3. $m_{UC} = 2f\overline{m}/\chi^2_{2f,1-\alpha/2}$
4. $R_{LC} = \exp[-(T/m_{LC})]$
5. $R_{UC} = \exp[-(T/m_{UC})]$
6. $g = 1 - \alpha = 1 - 2(\alpha/2)$

These formulas have been solved inversely where necessary for the various values and formulas used in the example below.

Example 8.17 Solve Example 8.6 (derived from Example 6.6) for the four double confidence limit questions, assuming exponentiality. The sample size is six, there were six failures, the mean is 76.8 hr, and the standard deviation is 14.6 hr. (1) Find the mean life for a 90% reliability, a 95% level of confidence, and a 40 hr mission time. (2) Find the mission limits for a 90% reliability and a confidence level of 95%. (3) Find the confidence level for a reliability of 90% and a mission time of 40 hr. (4) Find the reliability limits for a confidence level of 95% and a mission time of 40 hr.

Solution:

(1) The new mean life, $\overline{m}$, is
$\chi^2_{2(6),0.05/2} = 24.383$ (Table A3)
$m_{LC} = T/\ln(1/R) = 40/\ln(1/0.90) = 380$ hr
$\overline{m} = \chi^2_{2(6),0.05/2}m_{LC}/2f = 24.383(380)/2(6) = 772$ hr

In order to be 95% confident that no more than 10% of the items will fail before 40 hr, the product must be produced to an average mean life of 772 hr, assuming double limits. Note the difference between this mean life and the one determined with single-limit procedures (772 as opposed to 666 hr).

(2) The mission time limits, T_{LC} and T_{UC}, are
$\chi^2_{2(6),0.05/2} = 24.383$ (Table A3)
$\chi^2_{2(6),1-0.05/2} = 4.353$ (Table A3)
$m_{LC} = 2f\overline{m}/\chi^2_{2(6),0.05/2} = 2(6)(76.8)/24.383 = 37.8$ hr
$T_{LC} = m_{LC}[\ln(1/R)] = 37.8[\ln(1/0.90)] = 3.98$ hr
$m_{UC} = 2f\overline{m}/\chi^2_{2(6),1-0.05/2} = 2(6)(76.8)/4.353 = 212$ hr
$T_{UC} = m_{UC}[\ln(1/R)] = 212[\ln(1/0.90)] = 22.3$ hr

If the product is produced to an average mean life of 76.8 hr, we are 95% confident that no more than 10% of the units will fail before 3.98 hr nor after 22.3 hr (or 90% will fail between 3.98 and 22.3 hr). Compare this result to Example 8.6 where the single lower limit was 4.6 hr.

(3) The confidence level, g, is
$m_{LC} = T/\ln(1/0.90) = 40/\ln(1/0.90) = 380$ hr
$\chi^2_{2(6),\alpha/2} = 2f\overline{m}/m_{LC} = 2(6)(76.8)/380 = 2.425$
$\alpha/2 = 1.0$ (Table A3)
$g = 1 - 2(\alpha/2) = 1 - 2(1.0) = 0.0\%$

Since there cannot be a negative percentage, g must equal 0.0 instead of -1.0.

We have no confidence at all that more than 10% of the product (for a reliability of 90%), when produced to an average mean life of 76.8 hr, will last longer than 40 hr. If this distribution really were exponential, the quality of the product must be drastically improved to meet minimum specifications (a very costly proposal, usually). However, this distribution is NOT exponential. It is very important that the proper distribution be used.

(4) The reliability limits, R_{LC} and R_{UC}, are
$\chi^2_{2(6),0.05/2} = 24.383$ (Table A3)
$\chi^2_{2(6),1-0.05/2} = 4.353$ (Table A3)
$m_{LC} = 2f\overline{m}/\chi^2_{2(6),0.05/2} = 2(6)(76.8)/24.383 = 37.8$ hr
$m_{UC} = 2f\overline{m}/\chi^2_{2(6),1-0.05/2} = 2(6)(76.8)/4.353 = 212$ hr

$$R_{LC} = \exp[-(T/m_{LC})] = \exp[-(40/37.8)] = 34.7\%$$
$$R_{UC} = \exp[-(T/m_{UC})] = \exp[-(40/212)] = 82.8\%$$

If the product is produced to an average life of 76.8 hr, we can be 95% confident that no less than 34.7% nor more than 82.8% of the product will last beyond 40 hours.

Two-Sided Limits using the Normal

Normal confidence limit formulas are based on the following basic formulas (αa always refers to the reliability and αb to the chi-square):

1. $z_{\alpha a/2} = (T - m_{LC})/\sigma$ (if sample data are used, $\sigma = s$)
2. $z_{1-\alpha}a_{/2} = (T - m_{UC})/\sigma$
3. $m_{LC} = 2f\overline{m}/\chi^2_{2f,\alpha b/2}$
4. $m_{UC} = 2f\overline{m}/\chi^2_{2f,1-\alpha b/2}$
5. $g = 1 - 2(\alpha b/2)$
6. $R_{LC} = 1 - \alpha a/2$
7. $R_{UC} = 1 - (1 - \alpha a/2)$

These formulas have been solved inversely, as necessary, for the various formulas and values used in the example below.

Example 8.18 Solve Example 8.17 (Examples 6.6 and 8.6) for double-sided confidence limits using the normal. Since $\alpha a = 1 - 0.90$, $z_{\alpha a/2} = z_{0.05} = -1.645$ and $z_{1-\alpha a/2} = z_{0.95} = +1.645$ (from Table A1). Since $\alpha b = 1 - 0.95$, $\chi^2_{2f,\alpha b/2} = \chi^2_{2(6),0.05/2} = 23.549$ and $\chi^2_{2f,1-\alpha b/2} = \chi^2_{2(6),1-0.05/2} = 4.353$.

Solution:

(1) The new mean life, $\overline{m}$, is
$$m_{LC} = T - s(z_{\alpha a/2}) = 40 - 14.6(-1.645) = 64.0$$
$$\overline{m} = m_{LC}(\chi^2_{2f,\alpha b/2})/2f = 64.0(23.549)/2(6) = 126 \text{ fph}$$
(2) The mission time limits, T_{LC} and T_{UC}, are
$$m_{LC} = 2f\overline{m}/\chi^2_{2f,\alpha b/2} = 2(6)76.8/23.549 = 39.135$$
$$m_{UC} = 2f\overline{m}/\chi^2_{2f,1-\alpha b/2} = 2(6)76.8/4.353 = 212$$
$$T_{LC} = m_{LC} + s(z_{\alpha a/2}) = 39.135 + 14.6(-1.645) = 15.1 \text{ hr}$$
$$T_{UC} = m_{UC} + s(z_{1-\alpha a/2}) = 212 + 14.6(1.645) = 236 \text{ hr}$$
(3) The confidence level, g, is
$$m_{LC} = T - s(z_{\alpha a/2}) = 40 - 14.6(-1.645) = 64.0$$
$$\chi^2_{2(6),\alpha b/2} = 2f\overline{m}/m_{LC} = 2(6)76.8/64 = 14.396$$
$$\alpha b/2 = 0.306 \text{ (Table A3)}$$
$$g = 1 - 2(\alpha b/2) = 1 - 2(0.306) = 38.8\%$$
(4) The reliability limits, R_{LC} and R_{UC}, are
$$m_{LC} = 2f\overline{m}/\chi^2_{2f,\alpha b/2} = 2(6)76.8/23.549 = 39.135$$
$$m_{UC} = 2f\overline{m}/\chi^2_{2f,1-\alpha b/2} = 2(6)76.8/4.353 = 212$$
$$z_{\alpha a/2} = (T - m_{LC})/s = (40 - 39.135)/14.6 = 0.059$$
$$\alpha a/2 = 0.523 \text{ (Table A1)}$$

$$R_{LC} = 1 - sa/2 = 1 - 0.523 = 47.7\%$$
$$z_{1-\alpha a/2} = (T - m_{UC})/s = (40 - 212)/14.6 = -11.78$$
$$1 - sa/2 = 0.0 \text{ (Table A1)}$$
$$R_{UC} = 1 - (1 - \alpha a/2) = 1 - 0.0 = 100.0\%$$

Two-Sided Confidence Limits using the Weibull Graph

Weibull two-sided graph procedures will be illustrated using the following example.

Example 8.19 In a test of 10 units ($n = 10$), there were six failures during the test period (which lasted for three times the mission specifications). The failure times are listed in the following table. Use the Weibull graph to determine (1) the double-sided reliability limits for a mission of 100 hr and a confidence level of 90%; (2) the double-sided time limits for a reliability of 80% and a confidence level of 90%.

Solution: See Fig. 8.5 (see Table A5 for rank %).

Failure Times	50% rank	95% rank	5% rank
53	7	26	0.5
130	16	40	4
210	26	51	9
320	36	61	15
450	45	70	22
600	55	78	30

We are 90% confident that no less than 80% of the units will fail between 22.5 and 400 hr (Fig. 8.6).

At a mission time of 100 hr, we are 90% confident that the actual reliability falls between 65 and 97.3% (Fig. 8.6).

Double-Sided Limits using the Weibull Formulas

As in single-sided limits, the two Weibull parameters, B and G, must first be determined and then the answers to the four confidence limit questions can be found.

A. Find B and G (X is the time to failure and Y is the median rank values from Table A5 or one of the two formulas from Section 6.6).

$$B = \frac{[\Sigma\{\ln X)(\ln \ln[1/(1 - Y)]\}] - [(\Sigma \ln X)\{\Sigma \ln \ln [1/(1 - Y)]\}]/n}{\Sigma(\ln X)^2 - (\Sigma \ln X)^2/n}$$

$$a = \{\Sigma \ln \ln[1/(1 - Y)]\}/n - B[\Sigma \ln X]/n$$

$$G = \exp[(-0.0003277 - a)/B]$$

B. Find the four confidence limit values.

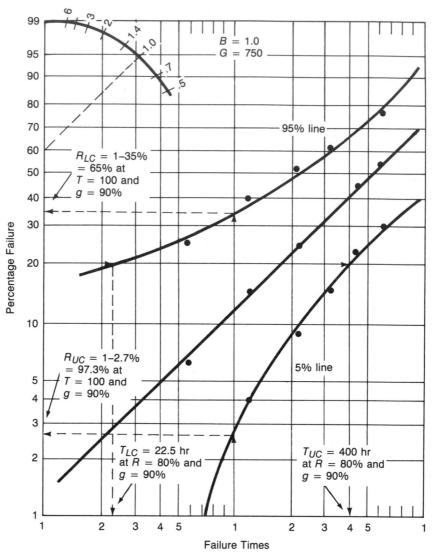

Fig. 8.6. Weibull double-sided confidence limits.

1. Find the new mean life, $\overline{m}$:
   $$G_{LC} = \exp\{\ln T - [\ln \ln(1/R)]/B\}$$
   $$\overline{m} = (\chi^2_{2f,\alpha/2})G_{LC}/2f$$
2. Find the mission time limits, T_{LC} and T_{UC}:
   $$G_{LC} = 2fG/\chi^2_{2f,\alpha/2}$$
   $$T_{LC} = G_{LC}(-\ln R)^{1/B}$$
   $$G_{UC} = 2fG/\chi^2_{2f,1-\alpha/2}$$
   $$T_{UC} = G_{UC}(-\ln R)^{1/B}$$

3. Find the confidence level, g:
$$G_{LC} = \exp\{\ln T - [\ln \ln(1/R)]/B\}$$
$$\chi^2_{2f,\alpha/2} = 2fG/G_{LC}$$
$$g = 1 - 2(\alpha/2) \text{ (Table A3)}$$
4. Find the reliability limits, R_{LC} and R_{UC}:
$$G_{LC} = 2fG/\chi^2_{2f,\alpha/2}$$
$$R_{LC} = \exp[-(T/G_{LC})^B]$$
$$G_{UC} = 2fG/\chi^2_{2f,1-\alpha/2}$$
$$R_{UC} = \exp[-(T/G_{UC})^B]$$

Example 8.20 Solve Example 8.6 for the four confidence limit questions for double limits. Input data are $R = 90\%$, $g = 95\%$, and $T = 40$ hr. The two Weibull parameters, from Example 6.6, are $B = 4.987$ and $G = 83.51$. $\chi^2_{2f,\alpha/2} = \chi^2_{2(6),0.05/2} = 23.549$ and $\chi^2_{2f,1-\alpha/2} = \chi^2_{2(6),0.95/2} = 4.353$.

1. Find the new mean life, $\overline{m}$:
$$G_{LC} = \exp\{\ln T - [\ln \ln(1/R)]/B\}$$
$$G_{LC} = \exp\{\ln 40 - [\ln \ln(1/0.90)]/4.987\} = 62.81$$
$$\overline{m} = (\chi^2_{2f,\alpha/2})G_{LC}/2f = (23.549)62.81/2(6) = 123 \text{ hr}$$
2. Find the mission time limits, T_{LC} and T_{UC}:
$$G_{LC} = 2fG/\chi^2_{2f,\alpha/2} = 2(6)(83.51)/23.549 = 42.555$$
$$T_{LC} = G_{LC}(-\ln R)^{1/B} = 42.555(-\ln 0.90)^{1/4.987} = 27.1 \text{ hr}$$
$$G_{UC} = 2fG/\chi^2_{2f,1-\alpha/2} = 2(6)(83.51)/4.353 = 230.2$$
$$T_{UC} = G_{UC}(-\ln R)^{1/B} = 230.2(-\ln 0.90)^{1/4.987} = 147 \text{ hr}$$
3. Find the confidence level, g;
$$G_{LC} - \exp\{\ln T - [\ln \ln(1/R)]/B\}$$
$$G_{LC} = \exp\{\ln 40 - [\ln \ln(1/0.90)]/4.987\} = 62.81$$
$$\chi^2_{2f,\alpha/2} = 2fG/G_{LC} = 2(6)(83.51)/62.81 = 15.9584$$
$$\alpha = 0.195 \text{ (Table A3)}$$
$$g = 1 - 2(\alpha/2) = 1 - 2(0.195) = 61.0\%$$
4. Find the reliability limits, R_{LC} and R_{UC}:
$$G_{LC} = 2fG/\chi^2_{2f,\alpha/2} = 2(6)(83.51)/23.549 = 42.555$$
$$R_{LC} = \exp[-(T/G_{LC})^B] = \exp[-(40/42.555)^{4.987}] = 48.0\%$$
$$G_{UC} = 2fG/\chi^2_{2f,1-\alpha/2} = 2(6)(83.51)/4.353 = 230.2$$
$$R_{UC} = \exp[-(T/G_{UC})^B] = \exp[-(40/146.6)^{4.987}] = 99.98\%$$

8.6 Sample Size and Parts per Million

When a product is to be tested to failure, it is seldom feasible or desirable to test very many. Destroyed units cannot be sold for profit. The Weibull testing procedures presented in the previous sections have presented a rule-of-thumb for these "continuous" data sample sizes. A rule often proposed for determining the sample size for "discrete" data derives from the formula for "binomial" confidence limits, as follows:

$$R_{LC} = (1 - g)^{1/n}$$

where

$$R_{LC} = \text{lower confidence limit on the reliability}$$

$$g = \text{confidence level}$$

$$n = \text{sample size}$$

The sample size formula can then be derived:

$$n = \ln(1 - g)/\ln R_{LC}$$

Remember that when a reliability is specified along with a confidence level, it is always a lower confidence limit.

Example 8.21 How many items need to be tested (failure free) to be 95% reliable at 90% confidence?

Solution:

$$n = \ln(1 - g)/\ln R_{LC} = \ln(1 - 0.9)/\ln 0.95 = 44.89 \text{ or } 45 \text{ items.}$$

Example 8.22 How many failure-free trips need to be made in a new car to be 98% confident that it is 99% reliable?

Solution:

$$n = \ln(1 - g)/\ln R_{LC} = \ln(1 - 0.98)/\ln 0.99 = 389 \text{ trips}$$

In the two preceding examples, the test hours per unit and miles per trip need to be determined separately. This emphasizes the inherent uncertainty in the preceding logic. Reliability *must* contain a time limit (mission time or T). The preceding formula does not contain a time limit; therefore, that formula is *not* a true reliability formula. It is, however, a fairly good measure of what the "discrete" sample size should be (as long as the hours, or miles, or actuations, etc., for each sample have been correctly determined elsewhere).

Parts per Million

Another error that has been propounded, to some extent, in reliability circles has to do with the concept of "parts per million," as it is frequently called. A low number of failures per million samples sounds like a high reliability (and it may well be). Once again, however, if it does not contain a time limit, it is *not* a reliability measure. It is just an excellent quality measure of percentage defect. Although a time limit can be assumed, thus converting the concept to a measure of reliability, the time limit is too often ignored in both application and calculation. (It is just too easily forgotten in the discussions and design and is not easily included in the calculations.) Unfortunately, this error has been perpetuated in some of the military standards on reliability.

8.7 Confidence Limits using the Binomial

Binomial confidence limit calculation procedures are identical to the exponential. A two-step procedure is used with one of the steps being the chi-square formula (the same chi-square formula used in the exponential). The only difference is that the second step formulas are binomial rather than exponential.

One-Sided Limits using the Binomial

The basic formulas are

$$m_{LC} = 2f\overline{m}/\chi^2_{2f,\alpha}$$

$$R_{LC} = (1 - 1/m_{LC})^T$$

These formulas have been solved inversely, where necessary, for the two-step solution procedures used in the following example.

Example 8.23 Solve Example 6.6 for the four confidence limit questions. The sample size is 6, the number of failures is 6, and the mean is 76.8. (1) Find the mean life for a 90% reliability, a 95% level of confidence, and a 40-hr mission time. (2) Find the mission time for a 90% reliability, a 95% level of confidence, and the preceding mean life, 76.8 hr. (3) Find the confidence level for a 90% reliability, a mission time of 40 hr, and a mean life of 76.8 hr. (4) Find the reliability for a confidence level of 95%, a mission time of 40 hr, and a mean life of 76.8 hr. Remember that the 40-hr desired mission time and the 90% desired reliability are actually lower limit specifications. In this problem, the 90% reliability specification will be a plain R and the 40-hr mission time specification will be a plain T.

Solution:

(1) The mean life, $\overline{m}$, is

$m_{LC} = 1/\{1 - \exp[(\ln R)/T]\}$
$m_{LC} = 1/\{1 - \exp[(\ln 0.90)/40]\} = 380$ hr
$\chi^2_{2(6),0.05} = 21.026$ (Table A3)
$\overline{m} = \chi^2_{2(6),0.05}(m_{LC})/2f = 21.026(380)/2(6) = 666$ hr

(2) The mission time, T_{LC}, is

$\chi^2_{2(6),0.05} = 21.026$ (Table A3)
$m_{LC} = 2f\overline{m}/\chi^2_{2(6),0.05} = 2(6)(76.8)/21.026 = 43.8$ hr
$T_{LC} = \ln R/[\ln(1 - 1/m_{LC})]$
$T_{LC} = \ln 0.90/[\ln(1 - 1/43.8)] = 4.56$ hr

(3) The confidence level, g, is

$m_{LC} = 1/\{1 - \exp(\ln R)/T]\}$
$m_{LC} = 1/\{1 - \exp[(\ln 0.90)/40]\} = 380$ hr
$\chi^2_{2(6),\alpha} = 2f\overline{m}/m_{LC} = 2(6)(76.8)/380 = 2.425$
$\alpha = 1.00$ (Table A3)
$g = 1 - 1.00 = 0.0\%$

(4) The lower reliability limit, R_{LC}, is

$\chi^2_{2(6),0.05} = 21.026$ (Table A3)

$m_{LC} = 2f\overline{m}/\chi^2_{2(6),0.05} = 2(6)(76.8)/21.026 = 43.8$ hr

$R_{LC} = (1 - 1/M_{LC})^T = (1 - 1/43.8)^{40} = 39.7\%$.

Two-Sided Limits using the Binomial

Binomial double-sided confidence limit formulas are based on the following six basic formulas:

1. $m_{LC} = 1/\{1 - \exp[(\ln R)/T]\}$
2. $m_{LC} = 2f\overline{m}/\chi^2_{2f,\alpha/2}$
3. $m_{UC} = 2f\overline{m}/\chi^2_{2f,1-\alpha/2}$
4. $R_{LC} = (1 - 1/m_{LC})^T$
5. $R_{UC} = (1 - 1/m_{UC})^T$
6. $g = 1 - \alpha = 1 - 2(\alpha/2)$

These formulas have been solved inversely where necessary for the various values and formulas used in the example below.

Example 8.24 Solve Example 8.6 (derived from Example 6.6) for the four double confidence limit questions, assuming the binomial. The sample size is 6, there were 6 failures, the mean is 76.8 hr, and the standard deviation is 14.6 hr. (1) Find the mean life for a 90% reliability, a 95% level of confidence, and a 40-hr mission time. (2) Find the mission limits for a 90% reliability and a confidence level of 95%. (3) Find the confidence level for a reliability of 90% and a mission time of 40 hr. (4) Find the reliability limits for a confidence level of 95% and a mission time of 40 hr.

Solution:

(1) The new mean life, $\overline{m}$, is

$\chi^2_{2(6),0.05/2} = 24.383$ (Table A3)

$m_{LC} = 1/\{1 - \exp[(\ln R)/T]\}$

$m_{LC} = 1/\{1 - \exp[(\ln 0.90)/40]\} = 380$ hr

$\overline{m} = \chi^2_{2(6),0.05/2}\, m_{LC}/2f = 24.383(380)/2(6) = 772$ hr

(2) The mission time limits, T_{LC} and T_{UC}, are

$\chi^2_{2(6),0.05/2} = 24.383$ (Table A3)

$\chi^2_{2(6),1-0.05/2} = 4.353$ (Table A3)

$m_{LC} = 2f\overline{m}/\chi^2_{2(6),0.05/2} = 2(6)(76.8)/24.383 = 37.8$ hr

$T_{LC} = \ln R/\ln(1 - 1/m_{LC})$

$T_{LC} = \ln 0.90/\ln(1 - 1/37.8) = 3.93$ hr

$m_{UC} = 2f\overline{m}/\chi^2_{2(6),1-0.05/2} = 2(6)(76.8)/4.353 = 212$ hr

$T_{UC} = \ln R/\ln(1 - 1/m_{UC})$

$T_{UC} = \ln 0.90/\ln(1 - 1/212) = 22.3$ hr

(3) The confidence level, g, is

$m_{LC} = 1/\{1 - \exp[(\ln R)/T]\}$

$m_{LC} = 1/\{1 - \exp[(\ln 0.90)/40]\} = 380$ hr

$\chi^2_{2(6),\alpha/2} = 2f\overline{m}/m_{LC} = 2(6)(76.8)/380 = 2.425$
$\alpha/2 = 1.0$ (Table A3)
$g = 1 - 2(\alpha/2) = 1 - 2(1.0) = 0.0\%$

Since there cannot be a negative percentage, g must equal 0.0 instead of -1.0.

(4) The reliability limits, R_{LC} and R_{UC}, are
$\chi^2_{2(6),0.05/2} = 24.383$ (Table A3)
$\chi^2_{2(6),1-0.05/2} = 4.353$ (Table A3)
$m_{LC} = 2f\overline{m}/\chi^2_{2(6),0.025} = 2(6)(76.8)/24.383 = 37.8$ hr
$m_{UC} = 2f\overline{m}/\chi^2_{2(6),0.975} = 2(6)(76.8)/4.353 = 212$ hr
$R_{LC} = (1 - 1/m_{LC})^T = (1 - 1/37.8)^{40} = 34.2\%$
$R_{UC} = (1 - 1/m_{UC})^T = (1 - 1/212)^{40} = 82.8\%$

Terms and Definitions

1. **Confidence limits.** Minimum and maximum acceptable values of the mean. The highest and lowest values in the confidence interval.
2. **Level of confidence.** The probability level associated with the confidence interval. The area under the normal curve between the confidence limits. Probability that the true mean lies between the confidence limits.
3. **Level of significance.** The probability that the true mean does not lie between the confidence limits.
4. **Chi-square.** A probability distribution used to calculate confidence limits.

Practice Problems

1. Ten parts were tested to failure. The failure times were 1, 6, 10, 20, 50, 70, 80, 85, 90, and 99 hr. What is the lower confidence limit for the reliability for a mission time of 10 hr and an α of 10%? What confidence can we have that the product will be 99% reliable at a mission time of 10 hr? What should the production mean life be for a 99% reliability at a confidence level of 90% and a mission time of 10 hr? What should the mission time specification be for a reliability of 99% and a 90% level of confidence? Solve for single-sided limits only and all four distributions.
   Answers:

	R_{LC} (%)	g (%)	$\overline{m}$ (hr)	T_{LC} (hr)
1. Exponential	75.7	0.0	1,413	0.36
2. Normal	75.0	3.9	141	−53.6
3. Weibull	68.7	0.0	10,624	0.06
4. Binomial	75.4	0.0	1,414	0.36

2. What must the production mean life be in order to be 99% confident that at least 95% of the computers being manufactured can operate for 24 hr without a failure, assuming (1) single limits? (2) double limits?

    Answers: 2155; 2634

3. Find the answers to Problem 2 for a mission time of 16 hr, a reliability of 90%, and a 95% confidence level.

    Answers: 455; 1851

4. One hundred and twenty components were run for 25 hr each during which 24 failed. The components were not repaired and replaced. What should the actual production mean life be for a 90% reliability, a 95% confidence level, and a 100-hr mission time? What should the mission time specification be? How confident can we be that the no more than 10% of the items will fail before 100 hours? If we wish to be 95% confident that the product will run for 100 hr without a failure, what must the reliability be? Use the distribution that applies and single limits.

    Answers: 1385; 8.124; 0.00; 27.3

5. Find the answers to Problem 4 for a mission time of 20 hr, a reliability of 90%, and a 95% level of confidence.

    Answers: 277; 8.124; 4.02; 77.2

6. The components of Problem 4 were continued on test until all failed. The total accumulated failure hours were 18,960. The standard deviation was 25 hr. Find the single-limit answers for the exponential and the normal.

    Answers:

	$\overline{m}$ (hr)	T_{LC} (hr)	g (%)	R_{LC} (%)
1. Exponential	1105	14.3	89.3	47.6
2. Normal	152	105	98.1	93.0

7. Find the answers to Problem 6 for a confidence level of 95%, a reliability of 99%, and a mission time of 50 hr.

    Answers:

	$\overline{m}$ (hr)	T_{LC} (hr)	g (%)	R_{LC} (%)
1. Exponential	14,902	0.53	3.11	38.8
2. Normal	182	78.8	50.5	93.0

8. A certain system was tested for 2800 hr. At the end of the test there had been 12 failures, which had been immediately repaired and replaced as they occurred. What must the production mean life be for a reliability of 90%, a 95% confidence level, and a mis-

sion time of 100 hr? What should the mission time specification be? What confidence can we have that the system will last longer than 100 hr? What is the reliability limit for a 95% confidence level and a mission time of 100 hr? Use the distribution that applies and single limits.

Answers: 1440; 16.2; 0.00; 52.2

9. Suppose the actual elapsed time of each failure in Problem 8 was 100, 400, 500, 700, 800, 900, 1100, 1200, 1300, 1500, 1800, and 2200 hr. Find the answers for the exponential, normal, and Weibull.

Answers:

	$\overline{m}$ (hr)	T_{LC} (hr)	g (%)	R_{LC} (%)
1. Exponential	1,440	72	63.2	86.5
2. Normal	1,327	−87.6	75.0	83.4
3. Weibull	859	144	99.9	93.7

10. Prove the answers to Example 8.13.

11. Prove the answers to Example 8.14.

12. Prove the answers to Example 8.15.

13. Prove the answers to Example 8.16.

14. Find the double limits for Example 8.13.

Answers:

	$\overline{m}$ (hr)	T_{LC} (hr)	T_{UC} (hr)	g (%)	R_{LC} (%)	R_{UC} (%)
1. Exponential	61.2	0.72	121	16.2	86.6	95.2
2. Normal	49.2	−15.4	69.7	17.2	74.9	98.4
3. Weibull	54.7	0.88	127.5	16.4	88.7	96.1

15. Find the double limits for Example 8.14.

Answers:

	$\overline{m}$ (hr)	T_{LC} (hr)	T_{UC} (hr)	g (%)	R_{LC} (%)	R_{UC} (%)
1. Exponential	31,114	11.5	4,770	0.0	33.4	67.9
2. Normal	1,014	242	1,372	39.0	54.9	99.9
3. Weibull	1,471	263	1,739	27.0	33.0	99.8

16. Find the double limits for Problem 6.

Answers:

	$\overline{m}$ (hr)	T_{LC} (hr)	T_{UC} (hr)	g (%)	R_{LC} (%)	R_{UC} (%)
1. Exponential	2,310	9.8	572	0.0	47.2	59.2
2. Normal	167	92.3	232	90.3	90.9	99.9

17. Solve Example 8.13 for single-sided limits using the binomial. The answers will be almost identical to the exponential.

18. Solve Example 8.14 for single-sided limits using the binomial. The answers will be almost identical to the exponential.

19. Solve Example 8.13 for double-sided limits using the binomial. The answers will be almost identical to the exponential.

20. Solve Example 8.14 for double-sided limits using the binomial. The answers will be almost identical to the exponential.

21. Solve Problem 1 for the mean life only, assuming no sample information is available and that the standard deviation and Weibull slope remain the same as determined in Problem 1. Compare to Problem 1 answers. Why are they different? (*Hint:* f now equals 1.)

    Answers:

Exponential = 2291; normal = 220; Weibull = 32,659; binomial = 2292

21. Solve Problem 9 for the mean life only, assuming no sample information is available and that the standard deviation and Weibull slope remain the same as determined in Problem 9. Compare to Problem 9 answers. Why are they different? (*Hint:* f now equals 1.)

    Answers:

    Exponential = 1422; normal = 425; Weibull = 1635

23. Solve Problem 2 for the normal if the standard deviation is assumed to be 5.
    Answers: 148; 199

24. Solve Problem 2 for the Weibull if the slope is assumed to be 2.
    Answers: 488; 623

25. Solve Problem 3 for the normal if the standard deviation is assumed to be 5.
    Answers: 67; 91

26. Solve Problem 3 for the Weibull if the slope is assumed to be 2.
    Answers: 148; 185

9 | MAINTAINABILITY AND AVAILABILITY

In this chapter the concepts of maintainability and availability are introduced. Maintainability provides a measure of the repairability of a system where a certain amount of failures can be allowed, while the availability provides a prediction of the probability that a complex system will be ready for use at any moment in time (will not be down for repairs).

Objectives

1. Have a knowledge of the nature of the maintenance function.
2. Understand the concept of maintainability, know the six optimization strategies, and apply the log-normal and other distributions to solve the three maintainability questions.
3. Use the log-normal and other distributions to determine maintainability confidence limits.
4. Understand the concept of availability and how it relates to maintainability and reliability and be able to solve availability problems.
5. Be able to calculate the number of spares needed to support the system reliability.

9.1 The Nature of Maintenance

In a system where a certain amount of failures can be allowed, the efficient repair and/or replacement of these failures is critical to the continued usefulness of the system. This repair and replacement of failures is called maintenance. Maintenance has a definite influence on operating costs, either through its own (maintenance) labor or through its effect on system downtime and efficiency. In reliability, maintenance can also be used to increase the probability that a system will continue to operate efficiently, given that it is allowed a certain amount of downtime for repairs.

Preventive Maintenance

The purpose of maintenance is to return a failed or deteriorating system to a satisfactory operating state. To do this, there are two extreme maintenance policies that can be applied. The first is to repair only when a component fails to operate (catastrophic or chance failure) or when its cost of operation becomes exorbitantly high (creeping or wearout failures). This is called corrective maintenance, or emergency repair. The second is to inspect periodically and then to repair and/ or replace as the need requires. This is called preventive maintenance (PM).

The purpose of PM is to eliminate the need for radical treatment sometime in the future (which is almost always much more expensive). PM, by its very nature, can be scheduled and controlled for a minimum cost; corrective action cannot. PM includes such things as inspecting, tuning, and major overhauls. In some types of systems (especially mechanical systems such as automobiles), PM can also include painting, waxing, chemical solution monitoring, and water treatment. The term PM is used broadly, but it is basically a philosophy of providing care that will protect and maintain the essential quality of the system.

Corrective Maintenance

When a failure occurs, it usually causes an emergency interruption of service, and this must be repaired and/or replaced immediately. This is usually a very expensive operation, certainly more costly than scheduled repairs. There are two types of corrective maintenance. The first type is called primary maintenance and is essentially the replacement of a failed part with a good one. Usually, the plug-in (or black box) concept is used where the whole section containing the failure is unplugged and a new section, or "black box," is plugged into the system to replace it. Thus, system downtime is not used to repair the actual failure. This can be done in the shop at a later time. The purpose of primary maintenance is to get the system up and running again as fast as possible. It is used in critical systems, such as aircraft, computers, missiles, etc., and where the time to locate and repair the failure is large.

The second type of corrective maintenance is called secondary maintenance. This type of maintenance usually consists of the follow-up repair (in the shop) of a failed plug-in; one that has already been replaced. Occasionally, this secondary maintenance can be done at the site of the failure, but only if the repair times are short, the system is not too critical, and downtime is not important.

Optimizing Maintainability

Considering the importance of maintenance, it is imperative to optimize it — to minimize both repair time and cost. There are six main strategies for doing this.

1. *Fault location and isolation.* This is frequently the most time-consuming task of all maintenance, especially with complex systems. This problem can be minimized by PM procedures where likely trouble spots are noted and logged, by built-in test equipment (push button line test, for instance) to locate deteriorating circuits, by simplicity in design of parts and components making trouble spots more easily visible, and by properly trained and experienced personnel.

2. *Repair time.* Trained personnel, design simplicity, proper tooling and repair facilities, and use of interchangeable and replaceable units are some of the more important means of reducing time to repair.

3. *Accessibility.* A part that is easier to get to not only reduces maintenance time, and therefore costs, but, perhaps more importantly, will minimize system downtime (reduce the time to get the system up and running again).

4. *Interchangeability.* This refers to plug-in devices where spares are instantly interchangeable with failed parts and to the ability of one unit to be used as replacement for several different components. This not only minimizes system downtime, but also reduces inventory costs and inventory stock-outs.

5. *Trained personnel.* With the increase in technology and system complexities, properly trained personnel are a must.

6. *Redundancy.* In large complex systems parallel components and subsystems can be built in to be used while the failures are repaired. (The parallel component can continue the service of the failed component so that the entire system can continue to operate while the failure is repaired.) Even entire systems can be used in parallel in this way to allow for more efficient maintenance actions.

9.2 Maintainability

Some systems by their very nature are unrepairable (aircraft in flight, missiles, etc.), but most can allow a certain number of failures within the mission time if provisions are made to repair them within a predetermined, specified, period of time. The measure of the ability of a system to be repaired within a designated period of time is called maintainability.

Mathematically, maintainability is defined as the probability of restoring a system (or component) to service in time D (where D, in this case, is the designated repair time). It can also be stated as the probability of repairing a failure in time D. This probability is represented by the area under the curve below the designated time, D (Fig. 9.1). In this respect, maintainability is the opposite of reliability, which is the area under the curve above the desired time limit.

For the distributions used in reliability, the formulas are essentially the same, except that the symbols T and $\overline{m}$ are changed to D and MTTR, respectively, and that maintainability is found by *not* subtracting from 1.00 ($M = \alpha$, while $R = 1 - \alpha$). These new maintainability symbols have meanings that are so similar to those used in reliability (T and $\overline{m}$) that the reliability symbols could have been used for maintainability. However, the new symbols are needed to reduce the possibility of confusion, especially when comparing and/or combining reliability and maintainability. D in maintainability is the desired repair time, the allowable repair time, or the maximum time that can be allowed to make a repair and to get the system up and running again. Since this definition is so close to the meaning of MTTR, the term "allowable downtime" will be used in its place, from now on, to minimize any possible confusion between the two terms. (The system is "down," of course, while a failure is being repaired.) MTTR in maintainability is the average time it actually takes to repair one failure (or the assumed time based on a similar product repaired in similar ways, or the assumed time based on system requirements, or a management assumption based on time, costs, etc.). Note that the upper case M and lower case m are different; they do *not* represent the same thing.

Maintainability data are always continuous; there is no such thing as a discrete repair time.

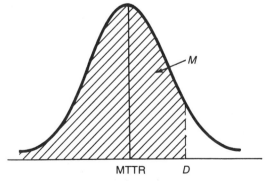

Fig. 9.1. Maintainability graph.

In maintainability, as in reliability, there are three questions that can be asked. They are:

1. What is the probability (maintainability or M) that a repair can be made within the specified downtime limit (D)?
2. What downtime (D) must be specified for a given maintainability (M)?
3. What is the new average repair time (MTTR) for a given M and D?

Note that question 3, as in reliability, does not need sample information for its calculation; all it needs is an assumed distribution. Thus, the necessary repair time (the average time in which each failure must be able to be repaired, or MTTR) for a given set of contract specifications can be determined even before the design is started. This can provide advance information to management for maintenance system design, repair tool specification and cost, training needed, manpower requirements, production downtime planning, etc. It is especially useful in providing advance cost information for contract negotiations and acceptance.

Most maintenance repair times are log-normally distributed, and so the log-normal distribution must be used, as well as the normal, exponential, and Weibull.

Maintainability Using the Log-Normal

Log-normal distributions are characterized by a positively skewed normal as shown in Fig. 9.2. In this figure, the times to repair are shown on the horizontal axis (the X axis). If the logarithms (to the base 10) of these times are plotted on the horizontal axis (instead of the actual time), the skew disappears and the plot becomes a normal (or almost normal) curve. Hence, the term log-normal (Fig. 9.3). Log-normal calculations are somewhat more complex than other distri-

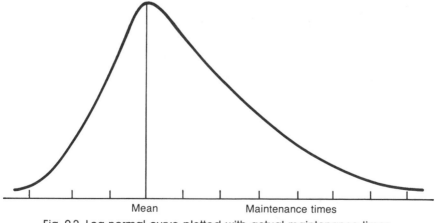

Fig. 9.2. Log-normal curve plotted with actual maintenance times.

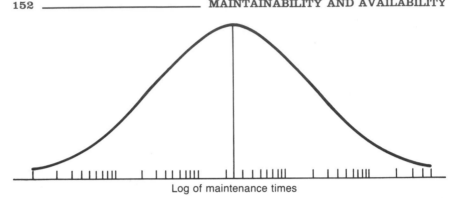

Log of maintenance times

Fig. 9.3. Log-normal curve plotted with the logarithms of actual maintenance times.

butions (except the Weibull), owing to the need to convert repair times into their respective logs.

The steps and formulas are:

A. Calculate the two parameters, a and s (t in the formulas is an individual repair time):

$$a = (\Sigma \ln t)/n$$

$$s = \sqrt{[\Sigma(\ln t)^2 - (\Sigma \ln t)^2/n]/(n - 1)}$$

B. Calculate the three maintainability values (formulas 2 and 3 are inverse solutions of the basic formula, formula 1):

(1) The basic formulas, for M, are
   $z_\alpha = (\ln D - a)/s$
   $M = \alpha$ (from Table A1)
(2) The inverse formula for D is
   $D = \exp[a + s(z_\alpha)]$
(3) The inverse formulas for MTTR are
   $a = \ln D - s(z_\alpha)$
   $\text{MTTR} = \exp[a + (s^2/2)]$

Example 9.1 The times to repair a certain system are as follows: 0.2, 0.6, 0.8, 1.1, 1.2, 1.3, 1.6, 1.9, 2.5 and 3.5 hr. (1) What is the maintainability for an allowed downtime (D) of 4 hr? (2) What must the allowed downtime (D) be for a maintainability of 90%? (3) What is the new MTTR for a 90% maintainability and an allowed downtime (D) of 4 hr?

Solution:

$$a = (\Sigma \ln t)/n = 1.4775/10 = 0.14775$$

$$s = \sqrt{[\Sigma(\ln t)^2 - (\Sigma \ln t)^2/n]/(n - 1)}$$

$$s = \sqrt{[6.054 - (1.4775)^2/10]/(10 - 1)} = 0.8052 \text{ (Table 9.1)}.$$

Table 9.1 Log-Normal Calculation Table

t	$\ln t$	$(\ln t)^2$
0.2	−1.60944	2.59030
0.6	−0.51083	0.26095
0.8	−0.22314	0.04979
1.1	0.09531	0.00908
1.2	0.18232	0.03324
1.3	0.26236	0.06883
1.6	0.47000	0.22090
1.9	0.64185	0.41197
2.5	0.91629	0.83959
3.5	1.25276	1.56941
$\Sigma t = 14.7$	$\Sigma \ln t = $ 1.4775	$\Sigma(\ln t)^2 = 6.05406$

(1) The maintainability, M, is

$z_\alpha = (\ln D - a)/s$

$z_\alpha = (1.386 - 0.14775)/0.8052 = 1.54$

$\alpha = 0.938$ (Table A1)

$M = \alpha = 93.8\%$

Given the preceding data, 93.8% of the failures can be repaired in 4 hr each, while 6.2% $(1 - 0.938)$ of the failures will each take longer than 4 hr to repair.

(2) The allowed downtime, D, that must be designated is

$z_{0.90} = +1.282$ (Table A1)

$D = \exp[a + s(z_{0.90})]$

$D = \exp[0.14775 + 0.8052(1.282)] = 3.25$ hr

Given the preceding data, we can expect no more than 10% of the failures to each take longer than 3.25 hr to repair (90% should take no longer than 3.25 hr to repair).

(3) The new mean repair time, MTTR, is

$z_{0.90} = +1.282$ (Table A1)

$a = \ln D - s(z_{0.90})$

$a = \ln 4 - 0.8052(1.282) = 0.354$

$\text{MTTR} = \exp[a + (s^2/2)]$

$\text{MTTR} = \exp[0.354 + (0.8052^2/2)] = 1.97$ hr

We must be able to average 1.97 hr per repair (instead of 1.4775) so that no more than 10% of the repairs will take longer than 4 hr. Obviously we are repairing faster than the specifications require, a very desirable situation. When confidence limits are added, this can, and often does, change.

Other Distributions

Although the log-normal is the distribution that most often applies to repair times, other distributions can sometimes be used. The exponential, normal, and Weibull can all be applied to maintainability. The procedures for applying these distributions to maintainability are identical to those used in reliability (Chapter 6). The formulas are also quite similar — the only differences being the substitution of $1 - M$ for R in all cases and a different interpretation of T and $\overline{m}$, as already explained.

MTTR, is calculated by dividing the sum of the repair times by the number of repairs, as follows:

$$\text{MTTR} = (\Sigma t)/n$$

where

MTTR $=$ the average time to repair a failure

t $=$ an individual repair time

n $=$ the number of repairs

Example 9.2 If the total time to repair 10 failures was 4.5 hr, what is the mean time to repair (MTTR)?

Solution:

$$\text{MTTR} = (\Sigma t)/n = 4.5/10 = 0.45 \text{ hr per failure}$$

The formulas for these three distributions are derived inversely, where necessary, from the basic formulas exactly as was done for reliability in Chapter 6. They are

1. Exponential
   $M = 1 - \exp[-(D/\text{MTTR})]$
   $D = \text{MTTR}[\ln(1/(1 - M)]$
   $\text{MTTR} = D/\ln[1/(1 - M)]$
2. Normal
   $z_\alpha = (D - \text{MTTR})/s$
   $M = \alpha$
3. Weibull
   First find the slope, B, and the characteristic life, G. (See Chapter 6 for the formulas and procedures.)
   $M = 1 - \exp[-(D/G)^B]$
   $D = G[-\ln(1 - M)]^{1/B}$
   $\text{MTTR} = \exp[\ln D - \{\ln \ln[1/(1 - M)]\}/B]$

As explained in Chapter 6, Section 6.6, the new mean repair time can be determined without recourse to sample information. Values can be assumed for MTTR and the standard deviation (or for B and G in the Weibull) and plugged into the formulas. In the problems in this chapter, MTTR, standard deviation, and B and G values already determined from the sample were used to calculate new mean repair time, but this was not necessary; any assumed values can be used. However, it is best to assume realistic values to allow for subsequent management analysis.

Example 9.3 Solve Example 9.1 for the exponential case. The sample MTTR is 1.47 hr/repair, the sample standard deviation is 0.968 hr, the desired maintainability is 90%, and the desired downtime is 4 hr.

Solution:

1. The new maintainability, M, is
   $M = 1 - \exp[-(D/\text{MTTR})]$
   $ = 1 - \exp[-(4/1.47)] = 93.4\%$
2. The new downtime, D, is
   $D = \text{MTTR}\{\ln[1/(1 - M)]\}$
   $ = 1.47\{\ln[1/(1 - 0.90)]\} = 3.38$ hr
3. The new mean repair time, MTTR, is
   $\text{MTTR} = D/\{\ln[1/(1 - 0.90)]\}$
   $\phantom{\text{MTTR}} = 4/\{\ln[1/(1 - 0.90)]\} = 1.74$ hr/repair

Example 9.4 Solve Example 9.1 using the normal. The sample MTTR is 1.47 hr/repair, the sample standard deviation (s) is 0.968 hr, the desired maintainability is 90% and the desired downtime is 4 hr. Since $\alpha = 90\%$, $z_\alpha = 1.282$ from Table A1.

Solution:

1. The new maintainability, M, is
   $z_\alpha = (D - \text{MTTR})/s = (4 - 1.47)/0.968 = 2.61$
   $M = \alpha = 0.995 = 99.5\%$ (Table A1)
2. The new downtime, D, is
   $D = \text{MTTR} + s(z_\alpha) = 1.47 + 0.968(1.282) = 2.71$ hr
3. The new MTTR is
   $\text{MTTR} = D - s(z_\alpha)$
   $\phantom{\text{MTTR}} = 4 - 0.968(1.282) = 2.76$ hr/repair

Example 9.5 Solve Example 9.1 using the Weibull graph. Only the new downtime and new maintainability can be determined using the Weibull graph.

Failure Times	50% Rank (Table A5)
0.2	7
0.6	16
0.8	26
1.1	36
1.2	45
1.3	55
1.6	64
1.9	74
2.5	84
3.5	93

Solution (see Fig. 9.5):

1. $D = 3.1$ hr
2. $M = 97\%$

Example 9.6 Solve Example 9.1 using the Weibull formulas. The desired maintainability is 90% and the desired downtime is 4 hr. The Weibull parameters, B and G, must first be determined before the three maintainability values can be calculated.

Solution:

A. The Weibull parameters, B and G, are

   1. $B =$
$$\frac{\{\Sigma(\ln x) \ln \ln[1/(1 - y)]\} - \{\Sigma(\ln x) \Sigma[\ln \ln[1/(1 - y)]]\}/n}{\Sigma(\ln x)^2 - (\Sigma \ln x)^2/n}$$
$$= 1.35$$

   2. $a = \{\Sigma \ln \ln[1/(1 - y)]\}/n - B[\Sigma \ln x/n] = -0.717$
   3. $G = \exp[(-0.00033 - a)/B] = 1.70$

B. The three maintainability values are as follows.

   1. The new maintainability, M, is
     $M = 1 - \exp[-(D/G)^B]$
     $= 1 - \exp[-(4/1.70)^{1.35}] = 95.8\%$
   2. The new downtime, D, is
     $D = G[-\ln(1 - M)]^{1/B}$
     $= 1.70[-\ln(1 - 0.90)]^{1/1.35} = 3.15$ hr
   3. The new MTTR is
     $\text{MTTR} = \exp\{\ln D - \ln \ln[1/(1 - M)]/B\}$
     $= \exp\{\ln 4 - \ln \ln[1/(1 - 0.90)]/1.35\} = 2.16$

Comparison of Results

A single example has been used throughout this chapter so that the various distributions can be compared. Note that there are major differences in their results to this problem. These differences will become even more pronounced when confidence limits are imposed. Obviously, the choice of the proper distribution is very important. If there is any doubt as to which distribution applies, it is probably best to use the Weibull, since it usually closely approximates the correct distribution, whatever it is. These comparisons are presented in Table 9.2.

Unrepaired Failures

The number of unrepaired failures is of interest to management (for planning purposes). This is the number of failures that are likely not to be repaired in time t. Time t in this case is any planning period that management chooses. The formula is

$$UF = (t/\text{MTBF})(1 - M) = (t/\overline{m})(1 - M)$$

A period of time, t, when divided by the average time between failures, gives the expected number of failures, on the average, that will occur during that time (t). $1 - M$ is the unmaintainability or the percentage of failures that will not be repaired in time. Multiplied together, they give the average number of unrepaired failures during each planning period. This calculation is independent of the method of determining M. The planning period (t), the allowed downtime (D), and the mission time (T) are also independent of each other (they do not have to be the same, or related to each other in any way).

Example 9.7 Calculate the number of unrepaired failures that would be expected in a planning period of one week (40 hr) for a 90% maintainability where the mean time between failures is 4 hr.

Solution:

$$UF = (t/\text{MTBF})(1 - M) = (40/4)(1 - 0.90) = 1.00 \text{ failure/week.}$$

9.3 Confidence Limits

As with reliability, maintainability requires a confidence limit to provide a reasonable assurance of accuracy. The previous analysis of maintainability was limited to the mean values, essentially a 50% level of confidence. Most people, and certainly most businesses, normally

Table 9.2 Comparison of Maintainability Values

Distribution	MTTR	D (hr)	M (%)
1. Log-normal	1.97	3.25	93.8
2. Exponential	1.74	3.38	93.4
3. Normal	2.76	2.71	99.5
4. Weibull	2.16	3.15	95.8

desire a much greater degree of confidence in the accuracy of their analysis results.

Unlike reliability, maintainability is limited to single-sided confidence limits. In reliability, some rationale can be developed for using double confidence limits (see Chapter 8, Section 8.5) even though the logic is weak and would seldom be used in real life. However, in maintainability this rationale simply does not apply. It makes no sense at all to specify a lower limit on the repair time. The faster the repair, the better.

As in reliability, there are four confidence limit questions that can be asked. They are:

1. What is the new MTTR that will be needed for a given confidence level (g), maintainability (M), and downtime (D)?
2. What is the necessary allowable downtime per failure (D) for a given confidence level (g) and maintainability (M)? The MTTR must also be used, but it can either be calculated from a test or be assumed.
3. What is the probability (the confidence level, g) that the real universe MTTR lies below a specified value (D) for a given maintainability (M)?
4. What is the maintainability limit (M_{UC}) for a given confidence level (g) and a given downtime (D)?

The same basic rationale and two-step application procedure developed for reliability confidence limits in Chapter 8 also applies to maintainability confidence limits. However, unlike reliability, a single-sided maintainability confidence limit is an upper limit. In the case of maintainability, the lower the better (we wish to keep our repair times as low as possible). A specified maintainability (M) with a confidence level (g) is automatically, therefore, an upper limit.

Figure 9.4 presents a picture of the maintainability confidence limit logic. The area under curve a below the upper repair time limit ($MTTR_{UC}$) represents the confidence level (g). The upper repair time limit ($MTTR_{UC}$) becomes the mean of distribution b. The area under curve b below the upper downtime limit (D_{UC}) represents the maintainability specification (M). The area under the curve below the specified downtime (D) represents the upper maintainability limit (M_{UC}). Note that the confidence level in maintainability is the area below the specified time, not above as in reliability, and is therefore equal to α ($M = \alpha$) (in reliability, $R = 1 - \alpha$).

Curve b represents the applicable distribution (log-normal, exponential, normal, or Weibull), while curve a represents the chi-square. Therefore, αa refers to the area under curve a, the chi-square curve, while αb refers to the area under curve b, the log-normal curve (or normal, or Weibull, or exponential, etc.). The confidence level and the maintainability now represent the area under the curve *below* the

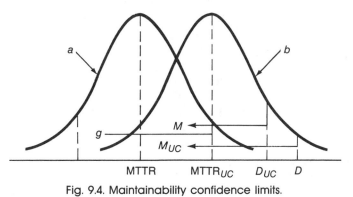

Fig. 9.4. Maintainability confidence limits.

critical value (not above as in reliability). Therefore, the tables are entered at the confidence level, not (1 − the confidence level), as in reliability. For a confidence level of 95%, for instance, the chi-square table (Table A3) would be entered as 0.95, not 1 − 0.95 as in reliability.

In each set of formulas, a MTTR confidence limit is required (MTTR$_{UC}$). This upper limit is used only as an intermediary value to get to the desired value. A MTTR limit is seldom used in real life and is, therefore, seldom required in the specifications. (The maximum downtime, D, is much more meaningful for control purposes.)

As in reliability confidence limits (see Chapter 8), all the distributions can be used with sample data or with assumed distribution parameters (an a and an s value for the log-normal; the mean only for the exponential; the mean and standard deviation for the normal; and the slope and characteristic life for the Weibull). Distribution parameters can be assumed by (1) comparing to a similar product produced by similar processes, (2) obtaining information from a long-term stable process where the product's data over a long time can be assumed to be the universe parameters, or (3) making an educated guess.

The chi-square is needed for calculating maintainability confidence limit values just as it was in reliability (Chapter 8). Therefore, the appropriate chi-square value must first be determined. The chi-square (χ^2) distribution provides a means for adjusting the limits for various sample sizes. As in reliability, the number of failures (number of repairs) is used instead of the sample size. The sample size in these distributions is just used to determine the degrees of freedom and, in maintainability, the degrees of freedom are determined by the number of repairs (which, of course, are in turn determined by the number of failures — only failures are repaired). When much sample information is available (many failures have been observed and their times to repair logged), the confidence interval can be narrow (the limits can be closer to the mean). The more information we have (the more failures we observe and measure), the more confidence we can have that our test results are accurate — that they correctly predict the real universe values.

Maintainability control limits, as in reliability (Chapter 8), can be thought of as a form of safety margin, with the margin equaling the difference between the MTTR and the lower limit ($MTTR_{UC}$) and the confidence level being the probability that this margin will not be exceeded. This safety margin must be large enough to be meaningful but not so large that costs become excessive. Using the chi-square in conjunction with the applicable distribution provides a solid probability estimate for this margin rather than just a simple guess. In addition, control limit values are automatically adjusted to fit the available information — the number of observed repairs. In this way, products with less failure/repair information (less observed failures/ repairs) are provided with greater safety margins.

If no test information is available (we are predicting from specifications only, for instance), a single failure and subsequent repair must be assumed, along with the distribution parameters already mentioned. Of course, the confidence interval is quite large (actually the maximum possible) when developed from only single failure/repair information. By making these assumptions, the formulas can be used to extrapolate design and maintenance information before a design is even started.

Maintainability Confidence Limits Using the Log-Normal

The following formulas have been derived from the basic formulas from Section 9.2 and the chi-square formula. Some of them have been derived by inverse solutions to these formulas. a and s are determined exactly as already explained in Section 9.2. a_C refers to an intermediate "calculated a" determined by the inverse use of the log-normal formulas, not from sample information.

1. To find the new MTTR from sample data:
   $MTTR_{UC} = \exp[\ln D - s(z_{ab})]$
   $a_C = MTTR_{UC}(\chi^2_{2f,\alpha a})/2f$
   $MTTR = \exp[\ln a_C + (s^2/2)]$
2. To find the downtime limit, D_{UC}:
   $MTTR_{UC} = 2f(e^a)/\chi^2_{2f,\alpha a}$
   $D_{UC} = \exp[\ln(MTTR_{UC}) + s(z_{ab})]$
3. To find the confidence level, g:
   $MTTR_{UC} = \exp[\ln D - s(z_{ab})]$
   $MTTR = \exp[a + (s^2/2)]$

$\chi^2_{2f,\alpha a} = 2f(\text{MTTR})/\text{MTTR}_{UC}$

$g = \alpha a$ (from Table A3)

4. To find the maintainability limit, M_{UC}:

$\text{MTTR} = \exp[a + (s^2)/2]$

$\text{MTTR}_{UC} = 2f(\text{MTTR})/\chi^2_{2f,\alpha a}$

$z_{ab} = [\ln D + \ln(\text{MTTR}_{UC})]/s$

$M_{UC} = \alpha b$ (from Table A1)

5. To find the new MTTR from assumed data, just use the formulas from 1 and substitute $f = 1$ (use the chi-square at two degrees of freedom).

Example 9.8 Solve Example 9.1 for the confidence limit values. The repair times were: 0.2, 0.6, 0.8, 1.1, 1.2, 1.3, 1.6, 1.9, 2.5, and 3.5 hr, respectively. Find: (1) The MTTR for a 90% maintainability (M), a 95% confidence level (g) and a downtime (D) of 4 hr. (2) The necessary allowable downtime limit (D_{UC}) for a 90% maintainability (M) and a 95% confidence level (g). (3) The confidence level (g) for a 90% maintainability (M) and a downtime (D) of 4 hr. (4) The maintainability limit (M_{UC}) for a confidence level (g) of 95% and a downtime (D) of 4 hr. Since $ab = 90\%$ and $\alpha a = 95\%$,

$z_{ab} = z_{0.90} = +1.282$ (from Table A1) and

$\chi^2_{2f,\alpha a} = \chi^2_{2(10),0.95} = 10.851$ (from Table A3)

From Example 9.1, $a = 0.14775$ and $s = 0.8052$.

Solution:

1. The new MTTR from the sample data ($f = 10$) is

$\text{MTTR}_{UC} = \exp[\ln D - s(z_{ab})]$

$\qquad = \exp[\ln 4 - 0.8052(1.282)] = 1.425$

$a_c = \text{MTTR}_{UC}(\chi^2_{2f,\alpha a})/2f$

$\qquad = 1.425(10.851)/2(10) = 0.773$

$\text{MTTR} = \exp[\ln(a_c) + (s^2/2)]$

$\qquad = \exp[\ln(0.773) + (0.8052)^2/2] = 1.07$ hr/repair

In order to be 95% confident that no less than 90% of the failures will take no more than 4 hr to repair, we must be able to average 1.07 hr per repair (*average* no more than 1.07 hr downtime per failure). We must improve our present average repair time from 1.47 hr per repair to 1.07 hr per repair. Note the difference a confidence level makes in the answer. In Example 9.1, part 3 (where no confidence level was specified) the MTTR was 1.97 hr per repair. Our product has now changed from acceptable to unacceptable. Where no confidence level is specified, this is equivalent to specifying a confidence level of 50%.

2. The downtime limit, D_{UC}, is
$$\text{MTTR}_{UC} = 2f(e^a)/\chi^2_{2f,\alpha a}$$
$$= 2(10)(e^{0.14775})/10.851 = 2.14$$
$$D_{UC} = \exp[\ln(\text{MTTR}_{UC}) + s(z_{ab})]$$
$$= \exp[\ln 2.14 + 0.8052(1.282)] = 6.0 \text{ hr}$$

For the preceding product (with the given test data), we can be 95% confident that no less than 90% of the failures will take no more than 6.0 hr to repair (no more than 10% will cause a downtime of more than 6.0 hr per failure).

3. The confidence level, g, is
$$\text{MTTR}_{UC} = \exp[\ln D - s(z_{ab})]$$
$$= \exp[\ln 4 - 0.8052(1.282)] = 1.425$$
$$\chi^2_{2f,\alpha a} = 2f(e^a)/\text{MTTR}_{UC}$$
$$= 2(10)(e^{0.14775})/1.425 = 16.27$$
$$\alpha a = 0.703 \text{ (Table A3)}$$
$$g = \alpha a = 70.3\%$$

We are 70.3% confident that no less than 90% of the failures will take no more than 4 hr to repair.

4. The maintainability limit, M_{UC}, is
$$\text{MTTR}_{UC} = 2f(e^a)/\chi^2_{2f,\alpha a}$$
$$= 2(10)(e^{0.14775})/10.851 = 2.14$$
$$z_{ab} = [\ln D - \ln(\text{MTTR}_{UC})]/s$$
$$= [\ln 4 - \ln 2.14]/0.8052 = 0.777$$
$$\alpha b = 0.781 \text{ (Table A1)}$$
$$M_{UC} = \alpha b = 78.1\%$$

We are 95% confident that no less than 78.1% of the failures will take no more than 4 hr to repair.

Example 9.9 Solve Example 9.8 for the new mean life only assuming that the a of 0.14775 and the s of 0.8052 were obtained from another, similar, product and that no sample information is available (therefore, $f = 1$). Since f now equals 1, the new chi-square value is

$$\chi^2_{2f,\alpha a} = \chi^2_{2(1),0.95} = 0.103$$

Solution:

$$\text{MTTR}_{UC} = \exp[\ln D - s(z_{ab})]$$
$$= \exp[\ln 4 - 0.8052(1.282)] = 1.425$$
$$a_c = \text{MTTR}_{UC}(\chi^2_{2f,\alpha a})/2f$$
$$= 1.425(0.103)/2(1) = 0.0734$$
$$\text{MTTR} = \exp[\ln(a_c) + (s^2/2)]$$
$$= \exp[\ln(0.0734) + (0.8052)^2/2]$$
$$= 0.102 \text{ hr per repair}$$

Note the difference between this answer and the one from Example 9.8 where sample information was available (1.07 to 0.102). When sample information is not available, repair capability must be improved dramatically in order to offset the greater uncertainty. This is almost certain to increase repair costs.

Maintainability Confidence Limits Using the Exponential

The following formulas have been derived from the basic formulas from Section 9.2 as well as the fundamental chi-square formulas. Many of the following formulas have been solved inversely from the basic formulas.

1. To find the new MTTR from sample data:
   $MTTR_{UC} = D/\ln[1/(1 - M)]$
   $MTTR = (MTTR_{UC})(\chi^2_{2f,\alpha})/2f$
2. To find the downtime limit, D_{UC}:
   $MTTR_{UC} = 2f(MTTR)/\chi^2_{2f,\alpha}$
   $D_{UC} = MTTR_{UC}\{\ln[1/(1 - M)]\}$
3. To find the confidence level, g:
   $MTTR_{UC} = D/\ln[1/(1 - M)]$
   $\chi^2_{2f,\alpha} = 2f(MTTR)/MTTR_{UC}$
   $g = \alpha$ (from Table A3)
4. To find the maintainability limit, M_{UC}:
   $MTTR_{UC} = 2f(MTTR)/\chi^2_{2f,\alpha}$
   $M_{UC} = 1 - \exp[-(D/MTTR_{UC})]$
5. To find the new MTTR from assumed data, just use the formulas from 1 and substitute $f = 1$ (use the chi-square at two degrees of freedom).

Example 9.10 Solve Example 9.8 using the exponential. MTTR is 1.47 hr, D is 4 hr, M is 90%, and g is 95%. Since α (g) is 95%, $\chi^2_{2f,\alpha} = 10.851$ (from Table A3).

Solution:

1. The new MTTR from sample data is
   $MTTR_{UC} = D/\ln[1/(1 - M)]$
   $= 4/\ln[1/(1 - 0.90)] = 1.74$
   $MTTR = (MTTR_{UC})(\chi^2_{2f,\alpha})/2f$
   $= 1.74(10.851)/2(10) = 0.94$ hr per repair
2. The downtime limit, D_{UC}, is
   $MTTR_{UC} = 2f(MTTR)/\chi^2_{2f,\alpha}$
   $= 2(10)(1.47)/10.851 = 2.71$
   $D_{UC} = MTTR_{UC}\{\ln[1/(1 - M)]\}$
   $= 2.71\{\ln[1/(1 - 0.90)]\} = 6.24$
3. The confidence level, g, is
   $MTTR_{UC} = D/\ln[1/(1 - M)]$
   $= 4/\ln[1/(1 - 0.90)] = 1.74$

$$\chi^2_{2f,\alpha} = 2f(MTTR)/MTTR_{UC}$$
$$= 2(10)(1.47)/1.74 = 16.90$$
$$\alpha = 0.665 \text{ (from Table A3)}$$
$$g = \alpha = 66.5\%$$

4. The maintainability limit, M_{UC}, is
$$MTTR_{UC} = 2f(MTTR)/\chi^2_{2f,\alpha}$$
$$= 2(10)(1.47)/10.851 = 2.71$$
$$M_{UC} = 1 - \exp[-(D/MTTR_{UC})]$$
$$= 1 - \exp[-(4/2.71)] = 0.772 = 77.2\%$$

Example 9.11 Solve Example 9.10 for the new MTTR only assuming that no sample information is available. D is 4 hr, g is 95%, and M is 90%. Since no sample information is available, f is assumed to be 1 and the $\chi^2_{2f,\alpha} = 0.103$ (from Table A3).

Solution:

$$MTTR_{UC} = D/\ln[1/(1 - M)]$$
$$= 4/\ln[1/(1 - 0.90)] = 1.74$$
$$MTTR = (MTTR_{UC})(\chi^2_{2f,\alpha})/2f$$
$$= 1.74(0.103)/2(1) = 0.090$$

Maintainability Confidence Limits Using the Normal

The following formulas have been derived from the basic formulas from Section 9.2 and the chi-square formula. Some of them have been derived by inverse solutions to these formulas. The standard deviation is shown as sigma (σ), but if sample data are used, the symbol should be s.

1. To find the new MTTR from sample data:
$$MTTR_{UC} = D - \sigma(z_{ab})$$
$$MTTR = (MTTR_{UC})(\chi^2_{2f,\alpha a})/2f$$
2. To find the downtime limit, D_{UC}:
$$MTTR_{UC} = 2f(MTTR)/\chi^2_{2f,\alpha a}$$
$$D_{UC} = MTTR_{UC} + \sigma(z_{ab})$$
3. To find the confidence level, g:
$$MTTR_{UC} = D - \sigma(z_{ab})$$
$$\chi^2_{2f,\alpha a} = 2f(MTTR)/MTTR_{UC}$$
$$g = \alpha \text{ (from Table A3)}$$
4. To find the maintainability limit, M_{UC}:
$$MTTR_{UC} = 2f(MTTR)/\chi^2_{2f,\alpha a}$$
$$z_{ab} = (D - MTTR_{UC})/\sigma$$
$$M_{UC} = ab \text{ (from Table A1)}$$
5. To find the new MTTR from assumed data, just use the formulas from 1 and substitute $f = 1$ (use the chi-square at two degrees of freedom).

Example 9.12 Solve Example 9.8 using the normal. MTTR is 1.47 hr, $s = 0.968$, D is 4 hr, M is 90%, and g is 95%. Since the standard deviation was derived from a sample, the symbol s was used instead of σ. Since $\alpha a = 95\%$ and $\alpha b = 90\%$:

$$z_{ab} = z_{0.90} = +1.282 \text{ (from Table A1), and}$$
$$\chi^2_{2f,\alpha a} = \chi^2_{2(10),0.95} = 10.851 \text{ (from Table A3)}$$

Solution:

1. To find the new MTTR from sample data:
$$\text{MTTR}_{UC} = D - s(z_{ab})$$
$$= 4 - 0.968(1.282) = 2.76$$
$$\text{MTTR} = (\text{MTTR}_{UC})(\chi^2_{2f,\alpha a})/2f$$
$$= (2.76)(10.851)/2(10) = 1.50 \text{ hr per repair}$$
2. To find the downtime limit, D_{UC}:
$$\text{MTTR}_{UC} = 2f(\text{MTTR})/\chi^2_{2f,\alpha a}$$
$$= 2(10)(1.47)/10.851 = 2.71$$
$$D_{UC} = \text{MTTR}_{UC} + s(z_{ab})$$
$$= 2.71 + 0.968(1.282) = 3.95 \text{ hr}$$
3. To find the confidence level, g:
$$\text{MTTR}_{UC} = D - s(z_{ab})$$
$$= 4 - 0.968(1.282) = 2.76$$
$$\chi^2_{2f,\alpha a} = 2f(\text{MTTR})/\text{MTTR}_{UC}$$
$$= 2(10)(1.47)/2.76 = 10.652$$
$$\alpha a = 0.954 \text{ (from Table A3)}$$
$$g = \alpha a = 95.4\%$$
4. To find the maintainability limit, M_{UC}:
$$\text{MTTR}_{UC} = 2f(\text{MTTR})/\chi^2_{2f,\alpha a}$$
$$= 2(10)(1.47)/10.851 - 2.71$$
$$z_{ab} = (D - \text{MTTR}_{UC})/s$$
$$= (4 - 2.71)/0.968 = 1.333$$
$$M_{UC} = \alpha b = 0.908 = 90.8 \text{ (from Table A1)}$$

Example 9.13 Solve Example 9.12 for the new MTTR only assuming that no sample information is available. D is 4 hr, g is 95%, and M is 90%. Since no sample information is available, f is assumed to be 1 and the $\chi^2_{2f,\alpha} = 0.103$ (from Table A3). Also the symbol σ is used for the assumed standard deviation (assumed to be the universe standard deviation) instead of the sample symbol s.

Solution:

$$\text{MTTR}_{UC} = D - \sigma(z_{ab})$$
$$= 4 - 0.968(1.282) = 2.76$$
$$\text{MTTR} = (\text{MTTR}_{UC})(\chi^2_{2f,\alpha a})/2f$$
$$= (2.76)(0.103)/2(1) = 0.142 \text{ hr per repair}$$

Maintainability Confidence Limits Using the Weibull Graph

Only the downtime limit (D_{UC}) and the maintainability limit (M_{UC}) can be determined using the graph.

Example 9.14 Solve Example 9.8 using the Weibull graph. D is 4 hr and M is 90%.

Solution:

Failure Times	95% Rank (Table A5)
0.2	0.5
0.6	3.7
0.8	9
1.1	15
1.2	22
1.3	30
1.6	39
1.9	49
2.5	61
3.5	74

From Figure 9.5:

1. The downtime limit, D_{UC}, = 3.1 hr.
2. The maintainability limit, M_{UC}, = 86%.

Maintainability Confidence Limits Using the Weibull Formula

As in Chapter 8, the following formulas are empirically derived (by trial and error, mostly) rather than determined from rigorous statistical and mathematical logic and, therefore, provide only approximations (although very close ones) to the more theoretically derived formulas. The theoretical formulas are much too complex for simple solution. Remember that confidence limit values are used to provide a form of safety margin in maintainability (as in reliability). Therefore, the following simplified formulas are quite adequate for this purpose. The Weibull parameters B and G are determined exactly as shown in Chapter 8 and in Section 9.2 of this chapter.

1. To find the new MTTR from sample data:
   $G_{UC} = \exp\{\ln D - \ln \ln[1/(1 - M)]/B\}$
   $\text{MTTR} = (G_{UC})(\chi^2_{2f,\alpha a})/2f$
2. To find the downtime limit, D_{UC}:
   $G_{UC} = 2f(G)/\chi^2_{2f\alpha a}$
   $D_{UC} = G_{UC}[-\ln(1 - M)]^{1/B}$

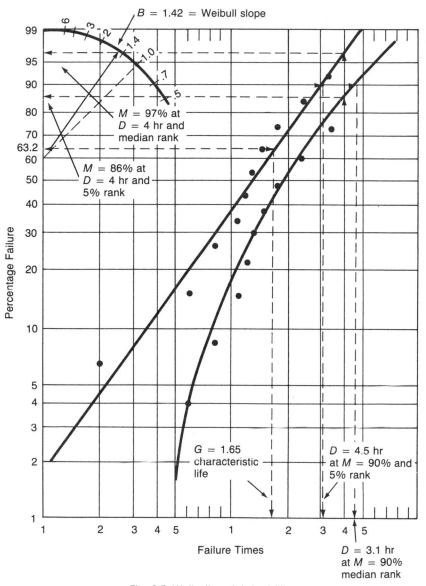

Fig. 9.5. Weibull maintainability.

3. To find the confidence level, g:
$$G_{UC} = \exp\{\ln D - \ln \ln[1/(1 - M)]/B\}$$
$$\chi^2_{2f,\alpha a} = 2f(G)/G_{UC}$$
$$g = \alpha \text{ (from Table A3)}$$

4. To find the maintainability limit, M_{UC}:
$$G_{UC} = 2f(G)/\chi^2_{2f,\alpha a}$$
$$M_{UC} = 1 - \exp[-(D/G_{UC})^B]$$

5. To find the new MTTR from assumed data, just use the formulas from 1 and substitute $f = 1$ (use the chi-square at two degrees of freedom).

Example 9.15 Solve Example 9.8 using the Weibull. $B = 1.35$ and $G = 1.70$ from Example 9.6, Section 9.2. In solving Weibull confidence limits from the formulas, the B and G parameters are always determined from the 50% (median) values only of Table A5. D is 4 hr, M is 90%, and g is 95%. Since $\alpha\alpha$ $(g) = 95\%$,

$$\chi^2_{2f,\alpha\alpha} = \chi^2_{2(10),0.95} = 10.851 \text{ (from Table A3)}$$

Solution:

1. The new MTTR from sample data is:
$G_{UC} = \exp[\ln D - \ln \ln[1/(1 - M)]/B\}$
$\quad = \exp\{\ln 4 - \ln \ln[1/(1 - 0.90)]/1.35\} = 2.16$
MTTR $= (G_{UC})(\chi^2_{2f,\alpha\alpha})/2f$
$\quad = 2.16(10.851)/2(10) = 1.17$ hr per repair

2. The downtime limit, D_{UC}, is
$G_{UC} = 2f(G)/\chi^2_{2f,\alpha\alpha}$
$\quad = 2(10)(1.70)/10.851 = 3.13$
$D_{UC} = G_{UC}[-\ln(1 - M)]^{1/B} =$
$\quad = 3.13[-\ln(1 - 0.90)]^{1/1.35} = 5.81$

3. The confidence level, g, is
$G_{UC} = \exp\{\ln D - \ln \ln[1/(1 - M)]/B\}$
$\quad = \exp\{\ln 4 - \ln \ln[1/(1 - 0.90)]/1.35\} = 2.16$
$\chi^2_{2f,\alpha\alpha} = 2f(G)/G_{UC}$
$\quad = 2(10)(1.70)/2.16 = 15.74$
$\alpha = 0.703$ (from Table A3)
$g = \alpha = 70.3\%$

4. The maintainability limit, M_{UC}, is
$G_{UC} = 2f(G)/\chi^2_{2f,\alpha\alpha}$
$\quad = 2(10)(1.70)/10.851 = 3.13$
$M_{UC} = 1 - \exp[-(D/G_{UC})^B]$
$\quad = 1 - \exp[-(4/3.13)^{1.35}] = 0.752 = 75.2\%$

Example 9.16 Solve Example 9.15 for the new MTTR only assuming that no sample information is available. D is 4 hr, g is 95%, and M is 90%. Since no sample information is available, f is assumed to be 1 and the $\chi^2_{2f,\alpha} = 0.103$ (from Table A3). B is still assumed to be 1.35.

Solution:

$G_{UC} = \exp\{\ln D - \ln \ln[1/(1 - M)]/B\}$
$\quad = \exp\{\ln 4 - \ln \ln[1/(1 - 0.90)]/1.35\} = 2.16$
MTTR $= (G_{UC})(\chi^2_{2f,\alpha\alpha})/2f$
$\quad = (2.16)(0.103)/2(1) = 0.111$ hr per repair

Comparison of Results

A single example has been used throughout this chapter in order that the various distributions can be compared. Note that there are major differences in their results to this problem. Obviously the choice of the proper distribution is very important. If there is any doubt as to which distribution applies, it is probably best to use the Weibull, since it usually closely approximates the correct distribution, whatever it is. These comparisons are presented in Table 9.3. Note the differences in these results when comparing them to the results when a confidence level was not required (Table 9.2). Table 9.2 shows that the product being manufactured is acceptable as to repair specifications (at least 90% can be repaired within the specified downtime of 4 hr). With a 95% level of confidence imposed, the nature of the problem changes to unacceptable (except for the normal).

Problem Analysis and Interpretation

The analysis and presentation format of the problems in this chapter follow the format of those presented in Chapters 6 and 8. As in these chapters, problems are answered for all questions at the same time (three maintainability and four confidence limit). In order to do this, especially in real life, some data may have to be assumed. For a more complete explanation and directions see Section 6.8, especially the paragraph headed "problem interpretation."

9.3 Availability

The probability that a system or equipment will be up and ready for use is called availability. Of course, in order to be ready for use, a system must either have not had a failure or, if a failure has occurred, have had it repaired. Thus, availability includes both reliability and maintainability. This leads to another definition of availability: The probability that a stated percentage of equipment or missions will have no downtime in excess of t in the mission time T. This definition implies that a poor reliability can be offset by a good maintainability. Indeed, good maintainability engineered into a system is one way of increasing the usefulness of that system.

Table 9.3. Comparison of Maintainability Confidence Limit Values

Distribution	MTTR ($f = 10$)	D_{UC}	g (%)	M_{UC} (%)	MTTR ($f = 1$)
1. Log-normal	1.07	6.00	70.3	78.1	0.102
2. Exponential	0.94	6.24	66.5	77.2	0.089
3. Normal	1.50	3.95	95.4	90.8	0.142
4. Weibull	1.18	5.76	74.0	75.3	0.111

Inherent Availability

A quick approximation of the availability of a system under ideal conditions where availability is potentially, at least, present in the design is given by the formula:

$$A = \frac{MTBF}{MTBF + MTTR}$$

Example 9.17 A system has a mean time between failures of 100 hr and a mean time to repair of 10 hr. What is the inherent availability?

Solution:

$$A = \frac{100}{100 + 10} = 0.9090 \text{ or } 90.9\%$$

9.4 Spares Requirements

If a system is allowed downtime for repairs, spare parts will inevitably be needed. The number of spare parts needed to maximize the probability that the system will be available is determined by the formula

$$\text{spares} = \lambda T + Z_{1-\alpha}\sqrt{\lambda T}$$

where

λT = the number of failures in the mission time T (equivalent to the mean)

$Z_{1-\alpha}$ = the standard normal deviate for $1 - \alpha$ (Table A1)

α = 1 − confidence level (determined by management)

$\sqrt{\lambda T}$ = standard deviation

Example 9.18 A system consists of 510 components. There are 10 component A's with a failure rate of 0.001 fph each. The other 500 component B's each have a failure rate of 0.0005 fph. The system is in use for 23 hr per day. Calculate the spares requirements for a 30-day period, if management wishes to be 99% confident that the spares will cover all the failures.

Solution:

$$T = 30(23) = 690 \text{ hr}$$

$$\lambda_A T = 0.001(690) = 0.69$$

$$\lambda_B T = 0.0005(690) = 0.345$$

$$\alpha = 1 - 0.99 = 0.01$$

$$Z_{1-\alpha} = 2.33$$

Component A

$0.69 + 2.33\sqrt{0.69} = 2.63$ spares needed for each of the 10 component A's

$10\,(2.63) = 26.3$ or 26 total component A's needed as spares for the 30 days.

Component B

$0.345 + 2.33\sqrt{0.345} = 1.71$ spares needed for each of the 500 component B's.

$1.71(500) = 855$ total spare component B's needed for the 30 days.

Terms and Definitions

1. **Maintenance.** The repair and/or replacement of failures.
2. **Reliability maintenance.** Increases the probability that a system will continue to operate efficiently, given that it is allowed a certain amount of downtime for repairs.
3. **Preventive maintenance.** Periodically inspect and then repair or replace as needed. Scheduled maintenance. Includes inspecting, adjusting, lubricating, replacing, tuning, and major overhauls.
4. **Corrective maintenance.** Repair when a component or system fails to operate (catastrophic or chance failure) or when its costs of operation become too high (creeping or wear-out failures).
5. **Primary maintenance.** Replacement of a failure using the plug-in concept.
6. **Secondary maintenance.** Repair of a failure in the shop after replacement.
7. **Plug-in replacement.** A means of swift replacement of a failure by means of prongs or other types of swift attachment techniques. Used in critical and expensive systems.
8. **Accessibility.** A technique designed to optimize maintainability by making all parts of the system easy to reach and easy to work on.
9. **Interchangeability.** A technique designed to optimize maintainability by designing similar parts so that they can be interchanged in the system.
10. **Allowable failures.** A management technique that increases reliability by providing for a certain amount of failure during system operating time.
11. **Maintainability.** The measure of the ability of a system to be repaired within a designated period of time. The probability of restoring a system to service in time (t).
12. **Mean time to repair.** Actual average time it takes to repair a failure. Designated by the symbol ϕ.
13. **Allowed time to repair.** Engineering or management determined maximum allowable time to repair any one failure. Designated by the symbol t.

14. **Mean repairs per time.** Actual average number of repairs that are made per time unit (usually 1 hr). Designated by the symbol μ. Reciprocal of mean time to repair.

15. **Availability.** The probability that a stated percentage of equipment or missions will have no downtime in excess of (t) during mission time (T). The probability that a system or component will be ready for use at any point in time.

16. **Spares requirements.** The number of spare parts, of each type, needed to support the desired availability.

Practice Problems

1. Failure repair data for a certain system is shown in the following table. For all four distributions, find: (a) Maintainability for an allowed downtime per failure of 9 hr, (b) Maintainability for an allowed downtime per failure of 9 hr. (c) The necessary downtime allowance for a maintainability of 95%. (d) The new mean (what the mean time to repair must be) for a maintainability of 95% and an allowed downtime of 9 hr

f = frequency of occurrence	Duration of each maintenance action in hours
1	1
2	2
3	3
5	4
7	5
10	6
8	7
4	8
3	9
1	10

Answers:

	M	D	MTTR
1. Exponential	79.0	17.3	3.00
2. Normal	94.3	9.1	−5.68
3. Log-normal	87.3	11.2	4.74
4. Weibull	89.7	10.1	5.93

2. Using the data from Problem 1 and assuming a desired downtime of 9 hr, a desired maintainability of 95%, and a desired confidence level of 90%, find: (a) MTTR, (b) D_{UC}, (c) g, and (d) M_{UC} for the exponential, normal, log-normal, and Weibull.

Answers:

	MTTR (hr)	D_{UC} (hr)	g (%)	M_{UC} (%)
1. Exponential	2.06	25.2	0.3	65.7
2. Normal	3.09	11.7	48.2	62.4
3. Log-normal	4.35	12.2	0.0	83.4
4. Weibull	4.07	14.6	33.7	57.1

3. Solve Problem 2 for the new MTTR only, assuming that no sample data are available. Also assume that the necessary distribution parameters (the standard deviation for the normal, the log-normal a and s, and the Weibull B) are identical to those already determined in Problem 2. Why are the answers different?
   (*Hint:* now $f = 1$.)
   Answers: 0.32; 0.60; 0.50; 0.55

4. A certain system has six failures that were repaired in 4, 6, 8, 12, 20, and 24 hr, respectively. For all four distributions find: (1) Maintainability for an allowed downtime of 21 hr. (2) The necessary downtime allowance for a maintainability of 99%. (3) The new mean time to repair (what must be specified) for a maintainability of 99% and an allowed downtime of 21 hr.
   Answers:

	M (%)	D (hr)	MTTR (hr)
1. Exponential	81.8	56.8	4.6
2. Normal	85.4	31.0	2.3
3. Log-normal	84.5	51.4	5.3
4. Weibull	83.3	39.5	7.6

5. Add a confidence level of 95% to the data from Problem 4 and solve for MTTR, D_{UC}, g, and M_{UC}.
   Answers:

	MTTR (hr)	D_{UC} (hr)	g (%)	M_{UC} (%)
1. Exponential	1.99	130	0.16	52.4
2. Normal	1.00	47.0	0.00	18.1
3. Log-normal	3.31	82.2	0.00	64.4
4. Weibull	3.29	91.0	3.46	40.4

6. Solve Problem 5 for the new MTTR only, assuming that no sample data are available. Also assume that the necessary distribution parameters (the standard deviation for the normal, the log-normal a and s, and the Weibull B) are identical to those already determined in Problem 5. Why are the answers different?

(*Hint:* now $f = 1$.)
  Answers: 0.24; 0.12; 0.27; 0.39

7. Find the inherent availability for the data in Problem 1 if the failure rate was (a) 0.1 failure per hour; (b) 0.01 failure per hour.
  Answers: 63.4%; 94.5%

8. Find the inherent availability for the data in Problem 4 if the mean life was (a) 40 hours per failure; (b) 120 hours per failure.
  Answers: 76.4%; 90.7%

9. Find the number of spare parts needed to support a mission of 2000 hr if the failure rate is (a) 0.1 failure per hour; (b) 0.01 failure per hour. A maximum of 5% stockout is desired.
  Answers: 223; 27

10. Solve Problem 11 for a maximum stockout of 1%.
  Answers: 233; 30

11. Find the number of spare parts needed to support a mission of 10 weeks at 5 days per week and 10 hr per day if the mean life is (a) 40 hours per failure; (b) 120 hours per failure. Use a maximum stockout of 2%. There are 100 of these units in service.
  Answers: 1,976; 836

10 | SAMPLING PLANS

In this chapter the concept and development of sampling plans are investigated. Sampling plans are used mostly to control incoming lots of finished product.

Objectives

1. Understand the concept of acceptance sampling including MRL, ARL, producer's risk, consumer's risk, LTFRD, and reliability risk.
2. Know the various types of sampling plans.
3. Be able to design a sampling plan from given ARL, LTFRD, R_α, and R_β values.
4. Understand the operating characteristic (OC) curve and be able to calculate one from a given sampling plan.
5. Know how to find the proper sampling plan from Mil-Std-105D.

10.1 Acceptance Sampling

The term acceptance sampling comes from quality control, where a lot is accepted or rejected based on inspecting a sample taken from the lot. The sample must be random and homogeneous. By random is meant that each unit in the lot has an equal chance of being chosen. Homogeneous means that the sample represents all of the classifications in the same ratio that they occur in the lot. Homogeneous also refers to a lot from one machine, one operator, one plant, etc. The reasons for sampling rather than 100% inspection are varied. The most important is the cost; it is usually much cheaper to sample. Often 100% inspection is impossible due to the size of the lot. If the tests are destructive, sampling is necessary or all units will be destroyed and none will be left to use. And, finally, psychological factors such as fatigue and boredom frequently cause a 100% inspection system to give more erroneous information than does a cheaper sampling plan. The ideal is a random sample of such size as to ensure maximum confidence at minimum cost.

The terminologies of acceptance sampling are somewhat unique. There are some new concepts involved as well as some new terms applied to old concepts. The most important are the following:

1. *MRL* means mean reliability level. This is just a new term given to the failure rate concept. It is the average failure rate (λ) for a group of samples. It is also called $\bar{r}$ in acceptance sampling and is found by using the appropriate failure rate formula (one of three already described to calculate λ). It can also be found by taking the reciprocal of the mean life ($\bar{r} = 1/\bar{m}$).

2. *ARL* means acceptable reliability level. This is a reliability requirement set by engineering and is usually expressed as failures per hour (or per 100 hr) (fph). This term should not be confused with the MRL, which is found from a group of samples (even though it also is expressed in failures per hour). ARL is also the allowed defective rate associated with producer's risk and is called r_1.

3. *Producer's risk* is the probability of rejecting a good lot and is usually expressed by α. This is *not* the same as the significance level (α) and should not be confused with it. Producer's risk is usually set at 5–10%. The rejection of a good lot is obviously an unnecessary cost to the producer, hence producer's risk. In reliability, producer's risk is called R_α and is defined as the probability that a time sample might be rejected even though the MRL $\leqslant$ ARL ($R_\alpha = 1 - R_s$).

4. *Consumer's risk* is the probability of accepting a bad lot and is usually denoted by β. The acceptance of a bad lot can have costly consequences to a consumer, hence consumer's risk. In reliability, consumer's risk is called R_β and is the probability of acceptance even when the reliability is worse than the poorest tolerable standard.

5. *LTFRD* means lot tolerance fraction reliability deviation. This is another defective rate limitation set by engineering, but is associated with consumer's risk rather than producer's risk and, when so defined, is called r_2. A good sampling plan will have the producer's and consumer's risks equal.

6. *A* is the acceptable number of failures in a sampling plan. If the sample runs for the whole allowed number of hours (T) with only *A* or less numbers of failures, the lot from which the sample is taken is accepted.

7. *Re* is the rejection number. If a sample has *Re* number of failures

before T hours are up, the lot is rejected. This Re should not be confused with the reliability R (which is expressed in percent rather than number of failures).

Reliability Risk

This subsection discusses one of the ways that the producer's reliability risk might be determined by engineering. This method illustrates the inherent relationships among MRL, A, and R_s.

Example 10.1 Find R_α given $A = 0$, $T = 2$, and MRL $= \bar{r} = 10/100 = 0.10$ fph.

Solution:

$$\lambda T = \bar{r}T = (0.1)(2) = 0.2$$

$$R_s = e^{-\bar{r}T} = e^{-0.2} = 0.819$$

$$1 - R_s = R_\alpha = 1 - e^{-\bar{r}T} = 1 - 0.819 = 0.18 \text{ or } 18\%$$

The producer's reliability risk (R_α) is the probability of rejecting good parts, which is a form of failure. Therefore, R_α is actually equal to $1 - P_s$ or P_f. In Example 10.1, the producer has an 18% chance of rejecting good lots. At $T = 4$, this probability increases to 33%, showing that the probability of rejecting a good lot will increase as the mission time T is increased. To decrease R_α, the acceptance number would have to be increased (at $C = 1$, $R_\alpha = 1.8\%$). Similar relationships exist for the consumer's risk R_β.

10.2 Sampling Plans

Fundamentally there are three types of sampling plans and five types of sampling methods. Each has its own peculiar advantages and disadvantages. A sampling plan must be chosen with care to maximize its advantages and minimize its disadvantages.

Types of Sampling Plans

1. *ARL plans.* These types of plans are used when the lot is homogeneous and discrete, as when coming from one machine or one process or delivered from one supplier. These plans are derived from the concept of acceptable quality level or AQL (ARL in reliability) and are designed to be used mostly when inspecting incoming lots from a producer. They were developed originally by the government and are imposed upon the producer by government procurement. Mil-Std-105D is the most important example of these types of plans. Although designed more for quality control, this standard can be adapted to reliability. However, more appropriate plans have now been developed

that apply directly to reliability (Handbook H108, for instance). ARL plans are the type that are used in reliability today.

2. *LTFRD plans.* These plans have been developed mostly for internal use. Since there are no outside requirements, usually, for these types of plans, they are designed mainly for minimization of costs, and thus cannot be applied to government purchasing.

3. *Average quality protection plans.* These plans are designed for use on lots that are mixed and of random size (mixed in storage, for instance). One hundred percent screening of bad product is usually required in these plans. The most famous of these plans is the Dodge–Romig tables. (Dodge and Romig were early pioneers in quality control.)

Sampling Methods

1. *Single sampling.* This method is used extensively because of its simplicity, ease of design, ease of use, and low administration costs. The method is to choose a sample of size n from the lot and inspect all units of the sample. If A or less defectives are found, accept the entire lot; otherwise reject the entire lot. Example: $n = 20$, $A = 3$, $Re = 4$. Choose a random sample of 20 from the lot and inspect all 20. If three or less defectives are found, accept the lot. If four or more are found, reject the lot.

2. *Double sampling.* As the name implies, two samples can (but do not have to be) drawn from the lot. Double and multiple sampling plans have the psychological advantage of a "second choice," since a second sample is drawn when the first one fails. These types of plans have the additional advantage of reducing inspection costs for good (and bad) product. If the product is consistently good (or bad), only one sample need be drawn, usually, for acceptance (or rejection). Since each sample of a multiple sample is always smaller than its single sampling counterpart, the costs are proportionately reduced. Of course, if the product is in between (neither good nor bad), the opposite effect will result. Both samples will usually be necessary in such a case and, since both samples together are larger than the one corresponding single sample, the costs are increased. Example: $n_1 = 15$, $n_2 = 15$, $A_1 = 1$, $A_2 = 3$, $Re = 4$. Select a random sample of 15 and inspect. If one or less defectives are found, accept the entire lot and stop sampling. If four or more defectives are found, reject the entire lot and stop sampling. If two or three defectives are found, take a second sample of 15 and inspect. If the total defectives from both samples is three or less, accept the lot, otherwise reject it.

3. *Multiple sampling.* These plans are identical to double sampling,

except that three or more samples are possible — up to seven in Mil-Std-105D.

4. *Sequential sampling*. This is an item-by-item sampling plan where the sample size is one and a decision (accept or reject) is made after each sample. In these plans, both R_e and A vary — they increase after each sample. Upward sloping limit lines are determined and graphed. After each sample, the cumulative number of defectives is graphed. When the cumulative defective line drops below the accept limit line, the lot is accepted. When it rises above the reject line, the lot is rejected. If neither happens, the test is usually halted at a predetermined number of samples and the lot is rejected. This is the plan most favored by reliability engineers. (See Fig. 10.1.)

5. *Continuous sampling*. Called CSP (CSP-1, CSP-2, etc.) plans, they are many and varied (Fig. 10.2). However, there are two main types — simple continuous sampling and continuous sampling allowing a defective. In CSP plans, 100% inspection is done until i continuous units are found free of defects. Then a fraction f of all units is inspected until one defect if found, after which 100% inspection is resumed and the cycle starts over. Variations usually allow one defect without returning to 100% inspection, although careful count of the sampled parts is kept until i more parts are found free of defects. Example: $i = 50$, $f = 1/10$. Inspect 100% until 50 units are found free of defects. Then inspect 1 of 10 until a defect is found, after which return to 100% inspection and start over. Example

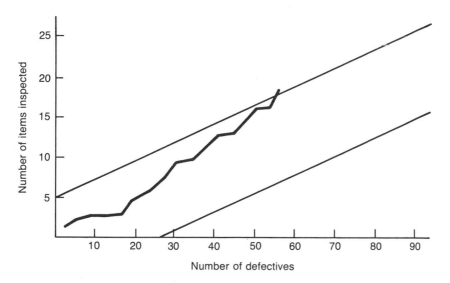

Fig. 10.1. Item-by-item sequential plan.

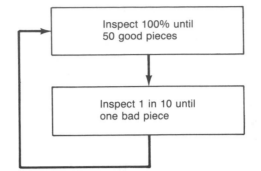

Fig. 10.2. Continuous sampling plan.

(see Fig. 10.3): $i = 50$, $f = 1/10$, $R = 2$. This is identical to the preceding example, except that when one defect is found, do not return immediately to 100% inspection, but continue to inspect 1 in 10. Keep count of amount inspected after the first defect. If another defect is found in the next 50 inspected, return to 100% inspection and start over. If not, continue to inspect 1 in 10, but discontinue counting.

Any of the preceding plans can be used for reliability, but the one favored by most practitioners is the sequential sampling plan.

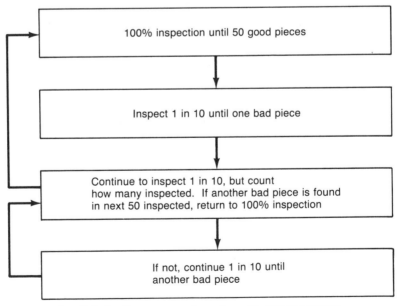

Fig. 10.3. Continuous sampling plan with allowed failure.

10.3 Designing a Plan

The purpose of the design of a sampling plan is to determine a test time (T). To design a sampling plan, four pieces of information must be given. First, the producer's risk (ARL $= r_1$) and the consumer's risk (LTFRD $= r_2$) must be determined by engineering. Remember that the producer's and consumer's risks are both failure rates and, as such, correspond to λ. Then, the acceptable levels of R_α and R_β must be determined by management. Since R_α is the probability of rejecting a good lot, it must be changed, in the calculations, to the probability of accepting a good lot ($1 - R_\alpha$) so that both R_α and R_β can be properly compared (both, then, are the probabilities of accepting). The next step is to calculate T for both producer's and consumer's risks using the inverse use of the Poisson (see Chapter 5) and using successively higher levels of c until the two T's match (are equal or as close to equal as they can get). Only when the two T's are equal is the plan considered to be fair to both producer and consumer. The reason for starting at $c = 0$ and working up is that the lowest possible c (acceptable number of failures) is desired.

Example 10.2

$$\text{ARL} = r_1 = 0.02 \text{ fph}$$

$$\text{LTRFD} = r_2 = 0.10 \text{ fph}$$

$$R_\alpha = 5\%$$

$$R_\beta = 10\%$$

There is a very good reason why the two failure rates, ARL and LTRFD, are quite different. Suppose that a consumer decides that more than 8% defective parts will cause undue problems during operation and 8% or less is acceptable. This then becomes the LTFRD. If the producer were to produce at this 8% level, 50% of his or her lots would be rejected. (Remember that 8% is only an average figure and that, on the average, 50% of the lots will be greater than 8% defective.) Therefore, the producer must produce at a much lower defective rate (ARL) in order to ensure that only a small percentage of the lots (R_α) will be rejected. The ARL can be calculated from this logic by using the normal curve techniques of Chapter 2.

Solution:

First trial $c = 0$:

$$e^{-r_1 T} = 1 - 0.05 = 0.95 = e^{-(0.02)T}$$

$$r_1 T = 0.05 \text{ (interpolate from Table A2)}$$

$$T_\alpha = r_1 T / r_1 = 0.05/0.02 = 2.5 \text{ hr}$$

$$e^{-r_2 T} = 0.10 = e^{-0.10T}$$

$$T_\beta = r_2 T / r_2 = 2.3/0.10 = 23 \text{ hr}$$

The two T's are not equal; try $c = 1$.

Second trial $c = 1$ or less:

At $c = 1$ or less and $1 - R_\alpha = 0.95$, $r_1 T = 0.36$ (from Table A2)

$$T_\alpha = r_1 T / r_1 = 0.36/0.02 = 18 \text{ hr}$$

At $c = 1$ or less and $R_\beta = 0.10$, $r_2 T = 3.9$

$$T_\beta = r_2 T / r_2 = 3.9/0.10 = 39 \text{ hr}$$

The two T's are still not equal; try $c = 2$.

Third trial $c = 2$ or less:

At $c = 2$ or less and $1 - R_\alpha = 0.95$, $r_1 T = 0.82$

$$T_\alpha = r_1 T / r_1 = 0.82/0.02 = 41 \text{ hr}$$

At $c = 2$ or less and $R_\beta = 0.10$, $r_2 T = 5.3$

$$T_\beta = r_2 T / r_2 = 5.3/0.10 = 53 \text{ hr}$$

The two T's are still not equal; try $c = 3$.

Fourth trial $c = 3$ or less:

At $c = 3$ or less and $1 - R_\alpha = 0.95$, $r_1 T = 1.35$ (Table A2)

$$T_\alpha = r_1 T / r_1 = 1.35/0.02 = 68 \text{ hr}$$

At $c = 3$ or less and $R_\beta = 0.10$, $r_2 T = 6.75$ (Table A2)

$$T_\beta = r_2 T / r_2 = 6.75/0.10 = 68 \text{ hr}$$

The sampling plan, then, is $T = 68$ hr and $A = 3$ failures. Test for 68 hr and repair and replace failures as they occur. If three or less failures occur by the end of the 68 hr, accept the lot. Whenever four failures occur (before the 68 hr are up), halt the test and reject the lot.

Suppose that successive trials produced no match between producer and consumer. This would become apparent when the sample size T_α for the producer becomes larger than the sample T_β for the consumer. At this point, there are four possible sample plans and a decision must be made as to which one will be used. None of the four will provide equal protection for both the producer and consumer, so the plan chosen must be the one that comes closest to this goal. This is, of

course, a management decision and is frequently determined by ne-
gotiation between producer and consumer.

Example 10.3 ARL = 0.01, LTFRD = 0.08, R_α = 5%, R_β = 10%.

Solution:

First trial c = 0:

$$e^{-r_1 T} = 1 - 0.05 = 0.95$$

$$r_1 T = 0.05 \text{ (interpolate from Table A2)}$$

$$T_\alpha = r_1 T/r_1 = 0.05/0.01 = 5 \text{ hr}$$

$$e^{-r_2 T} = 0.10$$

$$R_\beta T = 2.3 \text{ (interpolate from the Table A2)}$$

$$T_\beta = r_2 T/R_\beta = 2.3/0.08 = 29 \text{ hr}$$

The two *T*'s are not equal; try *c* = 1.

Second trial c = 1 or less:

At *c* = 1 or less and $1 - R_\alpha$ = 0.95, $r_1 T$ = 0.36 (Table A2)

$$T_\alpha = r_1 T/r_1 = 0.36/0.01 = 36 \text{ hr}$$

At *c* = 1 or less and R_β = 0.10, $r_2 T$ = 3.9

$$T_\beta = r_2 T/r_2 = 3.9/0.08 = 49 \text{ hr}$$

The two *T*'s do not match; try *c* = 2.

Third trial c = 2 or less:

At *c* = 2 or less and $1 - R_\alpha$ = 0.95, $r_1 T$ = 1.35

$$T_\alpha = r_1 T/r_1 = 1.35/0.01 = 135 \text{ hr}$$

At *c* = 2 or less and R_β = 0.10, $r_2 T$ = 6.75

$$T_\beta = r_2 T/r_2 = 6.75/0.08 = 84 \text{ hr}$$

Note that at *c* = 1, T_α was less than T_β, but at *c* = 2 this condition
is reversed. Any more trials at values in excess of *c* = 2 will only
find the T_α and T_β diverging at greater and greater speeds. Thus, one
of four plans must be chosen:

1. *c* = 1 and T_α = 36 hr (producer).
2. *c* = 1 and T_β = 49 hr (consumer).
3. *c* = 2 and T_α = 135 hr (producer).
4. *c* = 3 and T_β = 84 hr (consumer).

In choosing one of these four plans, one of four strategies could
be used.

1. Choose the plan with the lowest sample size. The plan is: $T_\alpha = 39$ hr and $c = 1$. This plan provides the lowest cost.
2. Choose the plan with the largest sample size. The plan is: $T_\alpha = 135$ hr and $c = 2$. This plan provides the lowest possibility of error.
3. Choose the plan that exactly meets the producer's stipulation and comes as close as possible to the consumer's stipulation. One of two plans must be chosen: (1) $T_\alpha = 36$ hr and $c = 1$ or (3) $T_\alpha = 135$ hr and $c = 2$. If plan (3) is chosen, the consumer's risk is then 0.1083 (3.9/36). If plan (2) is chosen, the consumer's risk becomes 0.05 (6.75/135). Since 0.1083 is closer to the consumer's stipulated risk of 0.08, choose plan (1) $T = 39$ hr and $c = 1$. (Test for 39 hr with repair and replacement. If there are one or less failures during the 39 hr, accept the lot; otherwise reject.)
4. Choose the plan that exactly meets the consumer's stipulation and comes as close as possible to the producer's. One of two plans must be chosen: (2) $T_\beta = 49$ hr and $c = 1$ or (4) $T_\beta = 84$ hr and $c = 2$. If plan (2) is chosen, the producer's risk becomes 0.0073 (0.36/49). If plan (4) is chosen, the producer's risk becomes 0.0161 (1.35/84). Since 0.0073 is closer to the producer's stipulated risk of 0.01, choose plan (2); $T_\beta = 49$ hr and $c = 1$.

Note: Another strategy is to average the sample sizes of the producer and consumer at the lowest c value of the four plans. This strategy, however, is technically incorrect. If used in Example 10.3, the plan would be $c = 1$ and $T = 42.5$ hr. This strategy could safely be used only in the case where the two times (T_α and T_β) are quite close in value.

10.4 Operating Characteristic Curve

An excellent tool for evaluating reliability plans is the operating characteristic (OC) curve. The OC curve can be used to determine the probability that a lot of a certain failure rate will be accepted or rejected. Figure 10.4 shows a typical OC curve. The ordinate (y axis) shows the probability that a lot will be accepted given a failure rate. The probability of rejecting the lot (P_f), then, is $1 - P_A$. Thus, in sample plan design, if $R_\alpha = 0.05$, the corresponding failure rate (r_1) would have to be found by subtracting 0.05 from 1 ($1 - 0.05 = 0.95$) locating 0.95 on the ordinate, moving to the right to the curve, and then reading the r_1 value from the abscissa (x axis). In Fig. 10.4 a lot with a failure rate of 0.02 would have a 75% chance of being accepted (given the sampling plan of $T = 89$ hr and $c = 2$). An example will serve to demonstrate how an OC curve is constructed.

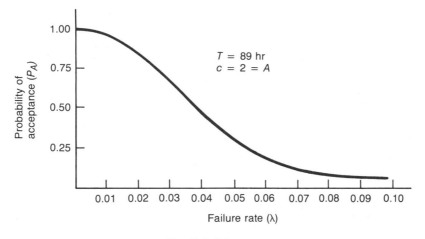

Fig. 10.4. OC curve.

Example 10.4 Using the information from Example 10.2, construct the OC curve ($T = 68$ hr and $A = 3$).

Solution:

With $T = 68$ hr and $c = 3$ or less, calculate a succession of points for the curve using $\lambda = 0.01, 0.03, 0.05, 0.07$, etc. ($P_A$ is the same as P_s.)

1. $\lambda - 0.01$; $\lambda T = 0.01(68) = 0.68$; $P_A = 0.995$ (Table A2).
2. $\lambda = 0.03$; $\lambda T = 2.04$; $P_A = 0.850$.
3. $\lambda = 0.05$; $\lambda T = 3.4$; $P_A = 0.558$.
4. $\lambda = 0.07$; $\lambda T = 4.76$; $P_A = 0.300$.
5. $\lambda = 0.09$; $\lambda T = 6.12$; $P_A = 0.140$.
6. $\lambda = 0.11$; $\lambda T = 7.48$; $P_A = 0.060$.
7. $\lambda = 0.13$; $\lambda T = 8.84$; $P_A = 0.025$.

The points represented by the seven λ's and their corresponding P_A's are now plotted on the graph to get the OC curve (Fig. 10.5).

10.5 Life and Reliability Standards

In the early days of reliability, the only sampling standards available were those in Mil-Std-105A (now Mil-Std-105D). However, this standard was specifically developed for quality control acceptance sampling of variables. Although Mil-Std-105D can be applied to reliability, there are now a host of standards more specifically designed for reliability.

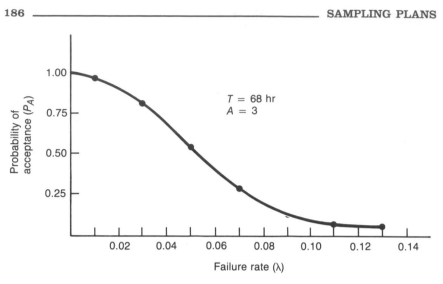

Fig. 10.5. OC curve.

Handbook H108 (procured from the federal government) is an excellent sample of the modern, more applicable standards.

Handbook H108 has many plans, but they can be classified into three main types:

1. Failure-terminated tests end when a predetermined number of failures occur. Acceptance is based on the number of text hours.
2. Time-terminated tests end at a predetermined time (number of hours). Acceptance is based on the number of failures.
3. Sequential tests are neither time nor failure fixed. Decisions depend on the accumulated results of the life test.

All three of these types can be with or without replacement, are based on the exponential distribution, and use mean life criterion (rather than failure rate). The plans are rather complex and each requires the use of several tables to apply.

Reliability sampling plans are much more difficult to determine than are comparable quality, or attributes, sampling plans. In attributes sampling, there is a definite relationship between lot size and sample size and thus the sample size can be determined easily from the lot size and the desired inspection level (see Mil-Std-105D). However, lot size is meaningless in reliability sampling where the sample is a number of hours rather than a number of discrete units. Because of this, sampling tables, such as Mil-Std-105D and others, can basically identify only a group of possible plans from which a single plan must be chosen. The decision as to which plan to use is based on a balance between cost and the producer's and/or consumer's risks. In MIL-HANDBOOK H108, these decisions are arbitrarily designed into the tables and are based on a desired sample size, a desired failure rate, or combinations of the two. An example using Mil-Std-105D will serve to illustrate the special nature of this problem.

Referring to Table 10.1, which is reproduced from Table IIA of Mil-Std-105D, the table is indexed by knowing the sample size and the desired quality level (called the AQL for acceptable quality level). The sample size is determined by reference to another table, which is in turn indexed by the lot size and the inspection level (both readily determined by engineering and/or management by examination of the nature of the problem and the cost and feasibility of the various lot sizes for that particular product). The AQL can be determined by management decision or estimated from calculations similar to those explained earlier in this chapter related to determination of the ARL.

Once the sample size and the AQL are determined, the sampling plan is simply read from the table. For instance, if a lot size of 600 and an inspection level of II (normal) are desired, the sample size code letter as determined from Mil-Std-105D is J. For an AQL of 1%, the sampling plan from Table 10.1 is: $n = 80$, $A = 2$. This means to take a random sample of 80 from the lot of 600 and inspect them. If two or less defective units are found, accept the lot. Otherwise reject the lot.

It should be noted at this point that the AQL can refer to either a percentage of defective parts or to a number of defects per 100 units. An AQL of 1.0 in Table 10.1, therefore, can refer to either 1% defective parts or to 1 defect per 100 units. When the chart is used to develop a reliability sampling plan, only the "number of defects per 100 units" aspect of the table is used.

In reliability, the AQL of Table 10.1 becomes the ARL and the "number of defects per 100 units" become the "number of failures per 100 units." The sample size is then the total test hours and the sample size code letter is ignored (since the code letter is determined by lot size, it cannot be used in reliability sampling where lot size is meaningless). Once a sample size is determined, it can be used as a test period for a single test unit or the total time can be divided among several units being tested simultaneously. (The number of test units would be determined by cost and/or time considerations.) A test of 500 hr, for instance, can be conducted, or 10 units for 50 hr each, etc.

Suppose that a certain product is desired to be operated in such a way that only one failure per 100 hr is expected. Using Table 10.1

Table 10.1. Single Sampling Plan (from Mil-Std-105D)

Acceptable Quality Levels (normal inspection)

Each acceptable quality level cell gives the pair **Ac Re** (Ac = acceptance number, Re = rejection number). ↓ = Use first sampling plan below arrow. ↑ = Use first sampling plan above arrow.

Sample size code letter	Sample size	0.010	0.015	0.025	0.040	0.065	0.10	0.15	0.25	0.40	0.65	1.0	1.5	2.5	4.0	6.5	10	15	25	40	65	100	150	250	400	650	1000
A	2	↓	↓	↓	↓	↓	↓	↓	↓	↓	↓	↓	↓	↓	↓	↓	↓	0 1	1 2	2 3	3 4	5 6	7 8	10 11	14 15	21 22	30 31
B	3	↓	↓	↓	↓	↓	↓	↓	↓	↓	↓	↓	↓	↓	↓	↓	0 1	1 2	2 3	3 4	5 6	7 8	10 11	14 15	21 22	30 31	44 45
C	5	↓	↓	↓	↓	↓	↓	↓	↓	↓	↓	↓	↓	↓	↓	0 1	1 2	2 3	3 4	5 6	7 8	10 11	14 15	21 22	30 31	44 45	↑
D	8	↓	↓	↓	↓	↓	↓	↓	↓	↓	↓	↓	↓	↓	0 1	1 2	2 3	3 4	5 6	7 8	10 11	14 15	21 22	30 31	44 45	↑	↑
E	13	↓	↓	↓	↓	↓	↓	↓	↓	↓	↓	↓	↓	0 1	1 2	2 3	3 4	5 6	7 8	10 11	14 15	21 22	30 31	44 45	↑	↑	↑
F	20	↓	↓	↓	↓	↓	↓	↓	↓	↓	↓	↓	0 1	1 2	2 3	3 4	5 6	7 8	10 11	14 15	21 22	30 31	44 45	↑	↑	↑	↑
G	32	↓	↓	↓	↓	↓	↓	↓	↓	↓	↓	0 1	1 2	2 3	3 4	5 6	7 8	10 11	14 15	21 22	30 31	44 45	↑	↑	↑	↑	↑
H	50	↓	↓	↓	↓	↓	↓	↓	↓	↓	0 1	1 2	2 3	3 4	5 6	7 8	10 11	14 15	21 22	30 31	44 45	↑	↑	↑	↑	↑	↑
J	80	↓	↓	↓	↓	↓	↓	↓	↓	0 1	1 2	2 3	3 4	5 6	7 8	10 11	14 15	21 22	30 31	44 45	↑	↑	↑	↑	↑	↑	↑
K	125	↓	↓	↓	↓	↓	↓	↓	0 1	1 2	2 3	3 4	5 6	7 8	10 11	14 15	21 22	30 31	44 45	↑	↑	↑	↑	↑	↑	↑	↑
L	200	↓	↓	↓	↓	↓	↓	0 1	1 2	2 3	3 4	5 6	7 8	10 11	14 15	21 22	30 31	44 45	↑	↑	↑	↑	↑	↑	↑	↑	↑
M	315	↓	↓	↓	↓	↓	0 1	1 2	2 3	3 4	5 6	7 8	10 11	14 15	21 22	30 31	44 45	↑	↑	↑	↑	↑	↑	↑	↑	↑	↑
N	500	↓	↓	↓	↓	0 1	1 2	2 3	3 4	5 6	7 8	10 11	14 15	21 22	30 31	44 45	↑	↑	↑	↑	↑	↑	↑	↑	↑	↑	↑
P	800	↓	↓	↓	0 1	1 2	2 3	3 4	5 6	7 8	10 11	14 15	21 22	30 31	44 45	↑	↑	↑	↑	↑	↑	↑	↑	↑	↑	↑	↑
Q	1250	↓	↓	0 1	1 2	2 3	3 4	5 6	7 8	10 11	14 15	21 22	30 31	44 45	↑	↑	↑	↑	↑	↑	↑	↑	↑	↑	↑	↑	↑
R	2000	↓	0 1	1 2	2 3	3 4	5 6	7 8	10 11	14 15	21 22	30 31	44 45	↑	↑	↑	↑	↑	↑	↑	↑	↑	↑	↑	↑	↑	↑

for an AQL (ARL) of 1.0, nine different sampling plans are identified. These plans are summarized as follows:

Sample size	Accept	Reject
13	0	1
50	1	2
80	2	3
125	3	4
200	5	6
315	7	8
500	10	12
800	14	15
1250	21	22

The decision as to which plan to use would depend on balancing the consumer's and producer's risk as closely as possible and on keeping the cost as low as possible. Obviously, the sampling plan with the lowest possible sample size that still closely balances the two risks would be the most desirable.

Terms and Definitions

1. **Sampling.** The process of measuring a few items from a group of items and assuming information about the group from the sample measurements.
2. **Acceptance sampling.** The process of using information from a sample in making decisions (accept or reject) about the group from which the sample was drawn.
3. **MRL (mean reliability level).** The average failure rate determined from measuring a sample or group of samples. Sometimes called $\bar{r}$ or $\bar{\lambda}$, it is the reciprocal of mean life.
4. **ARL (acceptable reliability level).** Determined by engineering design requirements, it is the failure rate that can be accepted by a producer. Sometimes called r_1 or λ_1.
5. **Randomality.** Each unit in the group has an equal chance of being chosen for the sample.
6. **Homogeneity.** All different features of the group are represented in the sample and to the same ratio. A lot from one machine, one operator, etc.
7. **Producer's risk (R_α).** The probability of rejecting a good lot. ARL is associated with producer's risk in devising a sampling plan.
8. **LTFRD.** Lot tolerance fraction reliability deviation. The failure rate that can be accepted by a consumer.

9. **Consumer's risk (R_β).** The probability of accepting a bad lot. Associated with LTFRD in developing a sampling plan.

10. **Unbiased sampling plan.** Producer's and consumer's protections are equal.

11. **Lot quality protection AQL plans.** A sampling plan for homogeneous, discrete lots. Usually consumer (government) imposed on the producer. Most reliability sampling plans are of this type.

12. **LTPD plans.** Sampling plans intended primarily for internal use (from department to department, for instance).

13. **Average quality protective plans.** Sampling plans for control of mixed lots (not homogeneous).

14. **Single sampling.** The use of one, single, sample in making acceptance decisions.

15. **Double sampling.** The use of one or two (if needed) samples in making acceptance decisions.

16. **Multiple sampling.** Using many samples (as many as seven) in making acceptance decisions.

17. **Sequential sampling.** An item-by-item sampling plan where a decision is made after each item is measured and the acceptance and rejection criteria vary after each item. Most used method in reliability.

18. **Continuous sampling.** An item-by-item method where a percentage is inspected after a certain amount is found free of defects, after 100% inspection. When a defect is found, the plan returns to 100% inspection and starts over.

Practice Problems

1. Find the total test time T and the acceptance criteria C when $r_1 = 0.02$ fph, $r_2 = 0.10$ fph, $R_\alpha = 0.05$, $R_\beta = 0.10$.
   Answers: 67 and 3

2. Find total test time and acceptance criteria for $r_1 = 0.01$ fph, $r_2 = 0.05$ fph, $R_\alpha = R_\beta = 0.05$.
   Answers: 190 and 4

3. Construct the OC curve for Problem 1.
4. Construct the OC curve for Problem 2.

11 | CONTROL CHARTS

The purpose of this chapter is to introduce some of the more important reliability charts that are used to control production.

Objectives

1. Understand the differences and similarities between sampling plans and control charts.
2. Understand how failure rate charts, for both variable and constant sample size, are used to control production. Be able to develop these failure rate charts, and their control limits, from sample data.
3. Understand the use of the sequential f chart and be able to develop this chart from sample data.

11.1 Sampling versus Control Charting

In Chapter 10, the concept and development of acceptance sampling plans was investigated. Acceptance sampling is used primarily to control incoming lots of finished product. Control charting, on the other hand, is used for control of the manufacturing process. As in acceptance sampling, control charts are based on sampling information. The difference is that control chart samples are taken from ongoing production (while the product is being produced), while acceptance sampling is done on the product after it is produced (and usually after it has been shipped). The decision objective in acceptance sampling is the acceptance or rejection of a lot. In control charting, the decision is concerned with the state of the production process, so that action can be taken immediately, while the product is being produced. These concepts apply equally to quality control and to reliability. However, in quality control the samples are units of production, or units of product, while in reliability the samples are time samples, units of time.

In order for control charting to be effective, the sample must be representative of the universe from which it comes; it must be random and homogeneous. By random is meant that each unit of production has an equal chance of being chosen for the sample. The sample will be homogeneous if the units come from a steady flow of product from a process that is under control. The process must also show a certain maturity of design. Maturity of design means that the units do not appreciably differ from each other, that all units are built to the same engineering specifications, and that all units are manufactured under the same conditions (using similar methods and processes). A final condition for effective control charting in reliability is that the failure rates be constant. Only when the failure rate is constant can the basic reliability formulas be applied with some degree of confidence.

There are four main advantages to control charting:

1. It provides quick information for corrective action, while the product is being produced.
2. It gives information on production trends, thus often allowing corrective action even before bad parts are run.
3. It gives information that can be used in subsequent acceptance sampling criteria.
4. If production trends are good and the failure rates are constant, acceptance sampling costs can often be decreased. (The control charts provide the confidence to reduce the acceptance sample size.)

There are several different types of sample control charts, but only the three most important will be presented here. The first is the variable failure rate chart, which plots the failures per hour when time samples (test periods) vary from one sample to another. The second chart is the constant failure rate chart, which plots failures per hour when time samples can be controlled (test periods are identical from sample to sample). Finally, sequential f charts plot cumulative failures against cumulative time samples. In the sequential f chart, the acceptance criterion varies with the time samples.

11.2 Control Charts for λ

This type of chart uses the failure rate (λ) to determine compliance to specifications. The purpose of this chart is to determine if the failure rates (λ) of the individual samples fail within the confidence limits of the ARL. The ARL (acceptable reliability level) is determined by engineering or is specified in the terms of the contract (negotiated between producer and consumer). If the ARL is not available, the MRL (mean reliability level) is used instead.

The ARL confidence limits are called control limits, and so a control limit is just a particular level of confidence limit (see Chapter 8) that is used as a standard. When a measurement falls within the control limits, the process is said to be in-control. Measurements that are in-control are considered to be close enough to the actual failure rate that they can be used as if they were the actual failure rate — the difference, if there is any, is not statistically significant. The Z value that determines these control limits is usually set at three, since ± 3 standard deviations give a confidence of over 99% that the real failure rate lies between the calculated control limits (99.73% for two-tailed limits).

There are two types of failure rate control charts — variable sample size and constant sample size. In reliability testing, the sample size is a time period (time sample). Reliability time samples are frequently of short duration, by necessity. The sample test must be continued until at least one failure occurs, or too many of the samples might end up with zero failure rate and thus invalidate the entire chart. When time samples can be extended so that there is a reasonable expectation that most of the samples will contain one or more failures, the constant sample size can be used. The great advantage of the constant sample size is that control limits can be determined ahead of time (based on a previous sample) and so production can be controlled more precisely (corrective action can be taken immediately, during the production run, rather than waiting until all the units are run). Another advantage of the constant sample size is that the control limits, upper and lower, are the same for all samples, making the chart much easier to use with less chance of error. Constant sample sizes should be used wherever possible.

Although the formulas calculate both upper and lower control limits, only the upper limit is of real importance in reliability testing. For shorter time samples, the lower limit usually calculates to less than zero. Since there cannot be a negative number of failures, the lower control limit in these cases is revised to zero. Zero failure rate, of course, would be the ideal state. For large time samples, the lower control limit will be a positive value, although usually close to zero. Although sample failure rates that fall below this lower control limit indicate exceptional quality, the process must still be considered as out-of-control and a search for an assignable cause must be instituted.

Sample failure rates that fall above the upper limit also indicate a process that is out-of-control. Assignable causes are assumed to be present and must be found and the process must be corrected. Of course, failure rates in excess of the upper control limits would be very undesirable, while those below the lower control limits would not; they would be very desirable indeed. The search for an assignable cause for these higher-level out-of-control points would have quite different objectives than the search for lower-level out-of-control causes. Lower-level out-of-control points show exceptional quality, and a search for the cause would be for the purpose of determining if it could be repeated on a regular basis. Higher-level out-of-control causes, on the other hand, would be determined for the purpose of making corrective action.

In both variable and constant sample size charts, the failures per hour must be calculated for each sample. The main difference between the two charts is the calculation of the control limits. For the variable chart, the control limits are calculated for each sample, while for the constant sample size, the control limits are calculated only once, for all samples (the limits are the same for all samples). The equations for both variable and constant sample sizes are the same:

$$CL = ARL \pm Z_{1-\alpha} \sqrt{\frac{ARL}{T}}$$

where

$$CL \quad = \text{control limits}$$

$$ARL \quad = \text{acceptable reliability level}$$

$$Z_{1-\alpha} \quad = \text{the standard normal deviate for } 1 - \alpha$$

$$1 - \alpha = \text{confidence level}$$

$$T \quad = \text{total test time (\textit{not} the test period)}$$

Note: When the ARL is not available, the MRL is substituted for the ARL in the above formula.

The use of the formula, for both variable and constant sample sizes, will be illustrated by the following examples.

Example 11.1 (Variable Sample Size) ARL = 1 failure/100 hr; $Z_{1-\alpha}$ = 3.0.

Sample Number	f	Hours	fph	Upper control limit (UCL)	Lower control limit (LCL)
1	1	100	0.0100	0.0400	0
2	2	210	0.0095	0.0307	0
3	3	120	0.0250	0.0374	0
4	3	300	0.0100	0.0273	0
5	1	75	0.0133	0.0446	0
6	6	200	0.0300	0.0312	0
7	3	275	0.0109	0.0281	0
8	6	300	0.0200	0.0273	0
9	1	125	0.0080	0.0368	0
10	1	80	0.0125	0.0435	0
11	1	95	0.0105	0.0408	0
12	13	375	0.0347	0.0255	0
13	1	110	0.0091	0.0386	0
Totals	42	2365			

Solution (The UCL and LCL in the preceding table are part of the solution): The control limit was calculated for sample number 1 as follows:

$$UCL = 0.0100 + 3\sqrt{\frac{0.01}{100}} = 0.400$$

$$MRL = 42/2365 = 0.0178 \text{ fph}$$

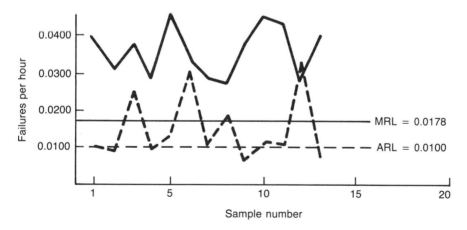

Analysis of the above chart shows that sample number 12 is above the upper control limit. This means that there is a 99.73% chance that something other than a chance cause (an assignable cause) is operating to cause the out-of-control condition. A search for an assignable cause must be instituted and, when found, action must be taken to correct the problem. The assumption that the process is out-of-control is further substantiated by the fact that the MRL is above the ARL (MRL = 0.0178 and ARL = 0.0100).

Note: The entire test (all 13 samples) is technically the sample with each of the 13 components a subsample. However, in actual practice, each subsample is just called a sample.

Example 11.2 (Constant Sample Size) ARL = 0.0100 fph; T = 200 hr; $Z_{1-\alpha}$ = 3.00

Subsample Number	f	Hours	fph
1	1	200	0.005
2	1	200	0.005
3	2	200	0.010
4	1	200	0.005
5	1	200	0.005
6	3	200	0.015
7	2	200	0.010
8	2	200	0.010
9	0	200	0.000
10	1	200	0.005
11	1	200	0.005
12	4	200	0.020
13	1	200	0.005
Totals	20	2600	

Solution:

$$\text{UCL} = \text{ARL} + 3\sqrt{\frac{\text{ARL}}{T}} = 0.0100 + 3\sqrt{\frac{0.0100}{200}}$$

$$= 0.0100 + 3\,(0.00707) = 0.0312$$

$$\text{LCL} = 0.0100 - 3(0.00707) = -0.0112 \text{ or } 0.00$$

$$\text{MRL} = 20/2600 = 0.0077 \text{ fph}$$

An analysis of this example shows that the process is in a state of control. Three conditions make this apparent. First, no one individual sample has a mean failure rate above the upper control limit. (In this example, it is impossible to have a mean failure rate below the lower limit.) If any one sample mean had exceeded the control limit, an assignable cause would have been assumed (something other than chance), a search would be instituted to find the cause, and corrective action would be taken to keep the condition from recurring in future

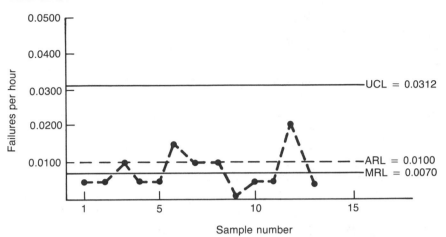

samples. Even if a sample mean is outside the control limits, this does not necessarily mean that the process, in general, is out of control. It may be that a variable just momentarily got out of hand and can be quickly and easily controlled once identified.

The second condition to look for that signals a process that is in control is the relationship of the MRL and the ARL. If the MRL is less than the ARL, as it is in this example, then the process is generally in control. This is true even if one or more of the samples are outside the control limits. If too many samples are out of control, of course, it would be impossible for the MRL to be lower than the ARL.

Finally, a trend analysis can be made to determine if the process, in general, is heading for trouble. If seven samples in a row are above the ARL, below the ARL, moving up or down, or moving alternately up and down across the ARL, the process can be considered to be out of control (or, at least, heading in that direction). Other trends that mean the same thing (there is a 3 sigma or 99.73% probability that a problem exists) are 9 out of 10 in a row and 11 out of 13.

11.3 The Sequential *f* Chart

The sequential *f* chart is an item-by-item test where decisions about acceptance or rejection are made after each failure. Thus, it can be used for both lot-by-lot acceptance sampling and for control chart sampling to control on-going production. Because a decision is made after each failure, the possibility exists of greatly reduced inspection costs. If the process is running exceptional quality (the fph are quite low), a decision to accept a lot or to continue the manufacturing process can be made quite early and testing time and costs can be reduced. Also, if the process is running bad quality (the fph are high), a decision to reject can be made quite early so that corrective action can be taken before too many bad parts are run. Only when quality is mediocre (neither high nor low) does the system equal most others in cost. Mediocre quality would cause a delay in making a decision, for failure after failure, until the sample is equal to that of single sampling plans. In fact, an additional criterion would have to be instituted in this case in order to truncate (stop) the test, otherwise it might go on forever without a decision point being reached.

The number of failures needed to reach a decision, in sequential plans, is based on products of conditional probabilities, using the binomial distribution. Thus, it is more generous, in its decision criteria, than are the single sampling plans. However, since reliability usually uses small sample sizes, the exponential can be used as an approximation to the binomial. Therefore, the sequential f chart is based on formulas using the exponential distribution.

The sequential f chart has no single failure criteria. Instead, two upward sloping parallel lines are computed in place of the rejection and acceptance numbers. Thus, acceptance and rejection criteria increase as the cumulative time sample increases. In order to calculate these lines (to determine their formulas), three values must first be calculated.

$$y_1 = \frac{\ln\,[(1 - R_\alpha)/R_\beta]}{\ln\,\{(r_2/r_1)\,[(1 - r_1)/(1 - r_2)]\}}$$

$$y_2 = \frac{\ln\,[(1 - R_\beta)/R_\alpha]}{\ln\,\{(r_2/r_2)\,[(1 - r_1)/(1 - r_2)]\}}$$

$$S = \frac{\ln\,[(1 - r_1)/(1 - r_2)]}{\ln\,\{(r_2/r_1)\,[(1 - r_1)/(1 - r_2)]\}}$$

Once these values are determined, they are substituted into the following straight-line formulas:

$$\text{Rejection line} \quad = f_R = ST - y_2$$

$$\text{Acceptance line} = f_A = ST - y_1$$

where

$y_1 =$ the origin of the acceptance line

$y_2 =$ the origin of the rejection line

$f_R =$ the number of unacceptable failures for a particular time sample

$f_A =$ the number of acceptable failures for any particular time sample

$S \;=$ slope of both lines

$T \;=$ the time sample

ln = the natural or naperian logorithm

It is obvious from these equations that certain values must be predetermined. As in acceptance sampling plans, the producer's risk (R_α), the consumer's risk (R_β), the ARL (r_1), and the LTFRD (r_2) must be given. These values are determined by engineering and management by techniques described in the previous chapter on acceptance sampling.

The two formulas above form two upward sloping parallel straight lines with the origin of the rejection line at y_2, the origin of the acceptance line at y_1, and an upward slope of S. (S is always positive.) Since the acceptance line starts at a negative number of failures (y_1 is always negative), it is obvious that acceptance must be deferred for a period of time (until the line crosses the abscissa or X axis).

The formulas are based on a single unit tested for the time required with failures being repaired and/or replaced as they occur. The chart is just as applicable to a group of units. The test period would then be reduced by a factor equal to the reciprocal of the group (10 items tested simultaneously, for instance, would reduce the test period to 1/10 that needed for one item). If one item needs a 100-hr test period, 10 items will need 1/10 of that or 10 hr each.

Example 11.3 Design a sequential *f* chart for ARL = r_1 = 0.01 fph, LTFRD = r_2 = 0.06 fph, R_α = 0.05, and R_β = 0.10

Solution:

$$y_1 = \frac{\ln\left[(1 - 0.05)/0.10\right]}{\ln\left\{(0.06/0.01)\left[(1 - 0.01)/(1 - 0.06)\right]\right\}}$$

$$= \frac{\ln\left[(0.95/0.10)\right]}{\ln\left[6\,(0.99/0.94)\right]}$$

$$= \frac{\ln 9.5}{\ln 6.31915} = \frac{2.25129}{1.84358}$$

$$= 1.22$$

$$y_2 = \frac{\ln\left[(1 - 0.10)/0.05\right]}{\ln\left\{(0.06/0.01)\left[(1 - 0.01)/(1 - 0.06)\right]\right\}}$$

$$= \frac{\ln\,(0.90/0.05)}{\ln\left[6\,(1.05319)\right]} = \frac{2.89037}{1.84358}$$

$$= 1.57$$

$$S = \frac{\ln\left[(1 - 0.01)/(1 - 0.06)\right]}{\ln\left\{(0.06/0.01)\left[(1 - 0.01)/(1 - 0.06)\right]\right\}} = \frac{0.05183}{1.84358}$$

$$= 0.0281$$

Rejection line $= f_R = ST - y_2 = f_R = 0.0281\ T + 1.57$
Acceptance line $= f_A = ST - y_1 = f_A = 0.0281\ T - 1.22$

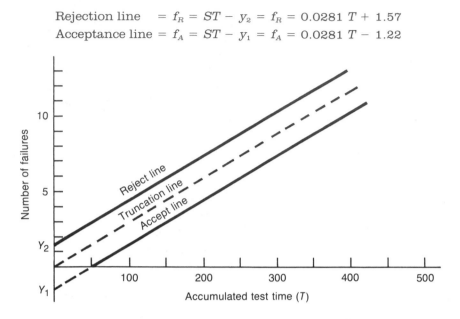

Analysis

1. The test must run for 43.416 hr before acceptance can be made $(f = 0 = 0.0281\ T - 1.22;\ T = 1.22/0.0281 = 43.416\ \text{hr})$.

2. The test cannot be rejected until 1.57 failures (two failures) $[f = 0.0281\ (0) + 1.57]$. At two failures, the test must run for 15.3 hr before rejection can be made $(f = 2 = 0.0281\ T + 1.57;\ T = (2 - 1.57)/0.0281 = 15.3\ \text{hr})$.

3. The acceptance and rejection lines are constructed by drawing a straight line between the two points determined by letting $T = 0$ and 100.

4. The truncation line, $f_t = 0.0281\ T$, is used to truncate (stop) a test of mediocre parts where no decision is reached after a certain pre-determined number of hours (predetermined by management and engineering). If at that time the number of failures is below the truncation line, the lot is accepted. If on or above, the lot is rejected. Assume, in the above example, that the test is to be truncated after 1000 hr is up (if no decision is reached before then). At that point, the lot would be rejected if there had been 28 failures or more (0.0281×1000).

5. Example 11.3 assumes one part tested for the requisite number of hours with repair and/or replacement of failures. Of course, more parts can be tested simultaneously for a lesser test period and still equal the required test time. For example, if 10 parts had been tested for 100 hr each, this would have been equivalent to one part tested for 1000 hr.

6. A different plan can be constructed at each sample size. For instance, at $T = 100$, the acceptance number is 2 [0.0281 (100) − 1.22], while the reject number is 4 [0.0281 (100) + 1.57]. At $T = 400$, the plan is: $T = 400$, $A = 10$, $R = 13$.

A more positive solution to Example 11.3 is to construct a rejection/acceptance table. The procedure is to calculate T, assuming subsequent levels of failures from 0 to 10 for both acceptance and rejection. Of course, the table can be extended beyond 10, if necessary.

Acceptance	Rejection	Time Sample
*	2	0– 15
0	3	16– 51
1	4	52– 86
2	5	87–122
3	6	123–158
4	7	159–193
5	8	194–229
6	9	230–264
7	10	265–300

An example will illustrate how the table is constructed (using an R of 3 as an example):

1. Calculate the maximum time sample of 51 hours [$T = (3 - 1.57)/0.0281 = 51$].
2. Calculate the acceptance number 0 [$f_A = 0.0281 (51) - 1.22 = 0.2131$].

Note: An * denotes no acceptance possible within that sample range.

Terms and Definitions

1. **Control chart.** A graphical representation of sampling data that supplies a picture of the status of the production process. Used to detect abnormal patterns of variations for the purpose of controlling production.
2. **Reliability sampling.** The period of time that a unit or system is under test.
3. **Maturity of design.** Refers to the point where the design of parts and processes is such that the product is produced with consistent quality.
4. **Failure rate chart.** A graphical representation of failure rates compared to statistical control limits. In a variable failure rate chart, the test period varies from sample to sample, while the test period does not change in a constant failure rate chart.

5. **Control limit.** Defines the limits of acceptable variation.
6. **Assignable cause.** A cause of out-of-control variation that is not due to chance alone.
7. **Sequential f chart.** A graphical method of analyzing failures where a decision is made after each sample and the decision criteria vary from sample to sample.

Practice Problems

1. Construct a failure rate control chart for the following data. Assume a confidence level of 99.73% (Z = 3) and the ARL = 2 failure per 100 hr. What is the MRL? Is the process in-control?

Subsample Number	f	Hours
1	2	220
2	3	350
3	3	170
4	4	200
5	6	250
6	1	80
7	1	90
8	1	110
9	4	350
10	8	360

2. Construct a failure rate control chart for the following data. The ARL = 2 failures per 100 hr, the confidence level is 99.73% (Z = 3), and the time sample is 300 hr. What is the MRL? Is the process in-control?

Subsample Number	f
1	0
2	1
3	0
4	1
5	5
6	7
7	2
8	8
9	6
10	4

3. Construct a sequential f chart for ARL = 0.002, LTFRD = 0.010, R_α = 0.05, and R_β = 0.10. What is the minimum number of test hours for acceptance? What is the minimum rejection number? Construct an acceptance/rejection table for f = 0 to 10.

12 | RELIABILITY AND DESIGN

In this chapter, the role of the design engineer in achieving reliability is examined. If reliability is properly designed into a system from the inception of the project, problems of production, operation, maintainability, and cost are minimized. Thus the importance of design in the realization of reliability is emphasized.

Objectives

1. Understand the goals and objectives of reliability as they apply to the design function.
2. Know some of the factors that must be used by design engineering in order to achieve the reliability objectives.
3. Understand the FMECA model.
4. Know the methods of estimating component reliability.
5. Know how reliability calculations can be used to guide the analysis and design of products.

12.1 Reliability Goals and Objectives

The general goal of reliability design is to see that the specified reliability is achieved and maintained. In order to do this, it first must be established just what is expected of the product. General statements are useless here; the requirements must be specific. (Words such as "more than" or "less than" are general descriptions that leave one still wondering what the goal actually is. What is "better than," for instance, and how do you measure it?) Therefore, statements of objectives and goals of reliability must include an actual probability level, a statement of acceptable operating conditions, and a time limit (the number of hours, days, months, etc., the product is expected to operate).

Objectives must be set for all stages of a product's evolution, for design, development, production, and operation. Some specific objectives would be:

1. The reliability of the product.
2. Various requirements relating to the product's expected performance.
3. Accuracy requirements.
4. The maintainability of the product — the probability that it can be repaired within a certain specified time limit. This includes ease of repair or replacement, ease of locating the failure, and cost of repairing the failure.
5. The operability of the product or its ability to be operated in the manner in which it was intended.
6. Producibility — the relative ease of manufacturing the product. This, of course, includes the cost of manufacture.
7. Procurability, which includes producibility as well as the whole procurement cycle of getting parts, materials, and labor together to produce the product and then shipping the product to the user. Time is the important factor in this objective.
8. The human factors involved with producing and operating the device must be considered; whether the physical and mental abilities of human beings are consistent with the producibility and operability requirements.
9. Expandability of the product refers to its ability to perform more functions, or more of the same or similar functions, by adding more parts. This objective may or may not be desirable depending on system requirements.
10. Versatility refers to the ability of the product to be used in several different systems to perform the same or related functions. This objective is usually highly desirable as it reduces the requirements for spare parts.

12.2 Reliability Design Factors

There are many diverse factors that assist in achieving the design objectives. Certainly the ability of the design engineer to recognize reliability principles and incorporate them into the design is most important. Ability in this area depends so much on the experience and training of the engineer. All design engineers should have some reliability training.

Some other important factors in reliability design are:

1. The product, and its various components, should be as simple as

possible without sacrificing any specification requirements. Complexity should be avoided.

2. Manual versus automatic operation and control. Automatic controls can be more reliable than can manual controls but may not have a human operator present at critical moments to make the type of decisions only humans can make. Design specifications should suggest which one is needed.

3. Speed versus reliability, replaceability, interchangeability, and maintainability. It may be more advantageous to keep a short-term system simple and speedily producible while a long-term system, needing a higher reliability, might require different design requirements in maintainability, interchangeability (less spares would be needed), and replaceability (quick changes in replacing failures).

4. Aptitude of operators versus the capability of the machine. Make sure the operators are physically capable of performing the required tasks. Consider if requirements exceed limits of human physical strength, that humans are not required to perform too many tasks too fast, that they are not required to assimilate too much information too fast, that information is delivered in a readily understandable form (clockwise instead of counterclockwise, high to low instead of low numbers on top, etc.), and other facts of human engineering design.

5. Technical and environmental factors. Consider such limitations as weight and size that might also limit accessibility and maintainability. Consider biological growth, sand and dust, humidity, heat and cold, and all other environmental factors that could affect design requirements.

6. Replaceable modules versus unit construction. Critical and high-cost systems would work better with replaceable modules (keep the system going with minimum delays), while low-cost, simple systems might be more advantageous as unit constructions.

7. Consider the effects of vibration, shock, and mechanical structure of the product.

8. Standard parts versus special design parts. Special design parts may be able to be made more reliable and useful for a specific purpose but standardized parts are readily available, less costly (usually), and require little or no design commitment.

9. Make sure the product can perform as required under all probable conditions of stress, power fluctuations, contamination, variations in operating conditions, age, etc.

10. Make sure that the interrelations between the electrical and the mechanical functions are optimum.

11. Consider problems with shock mountings, fasteners, wiring,

transportation, packaging, shipping, storage, identification markings, and installation.

12.3 Failure Modes Effects and Criticality Analysis

Probably the most useful tool that design engineers have today to ensure reliable products is the FMECA model. (FMECA is an acronym for Failure Modes Effects and Criticality Analysis.) FMECA is used to analyze equipment and systems to determine effects of individual components on the entire system and on each subsystem. First, major assemblies of the equipment are listed, after which each assembly is broken down into its component elements. Each component is then studied to see how it could fail, what could cause each type of failure, and the effect of this failure on other components, subassemblies, and the entire product. Failure rates can then be estimated and calculations can be made to determine the overall probability that the product will operate for a certain period of time, or the probable operating time between failures (predict the system reliability). Critical components are then identified and listed in the order of their criticality.

Critical components are those that affect safety, interfere with the mission goals, require special handling, increase maintenance loads to unbearable levels, have long lead times for procurement, have high replacement costs, etc. High-energy, high-pressure, and combustion systems are examples of systems that are usually critical by their very nature. Another fairly easily identified critical condition is the "single-point" component. These are parts that act similar to "bottleneck" operations in an assembly line. (In an electronic system, it is a single component through which all current paths must pass.) A failure here will almost always have a major impact on product performance. (Redundancy and fail safe devices are used to reduce these types of failure possibilities.)

Criticality is determined by multiplying the probability of damage by the probability of failure by the ratio of occurrence (of the critical mode to all possible failure modes). The various critical components are then listed with their failure modes, probability values, effects, and criticality ranking. The highest-ranked critical component — the most critical — is analyzed first, after which the others are analyzed in the order of their criticality.

One reason the FMECA model is so popular is that it maximizes engineering efficiency. Engineering effort is directed toward those failure modes that are most likely to affect product performance, and affect it the most. With FMECA, product reliability is maximized at the design stage (the least cost method of maximizing reliability), with corresponding minimization of engineering effort and costs. Most low-level criticality parts, for instance, may never have to be analyzed if product liability is properly increased (to an acceptable level) by a few high-level criticality parts.

12.4 Failure Rate Estimation

In order to properly and meaningfully predict reliability, the failure rate of the individual component parts must be known or must be estimated. In the "educated guess" method of estimating failure rates, an estimate of the proposed component is made based on the engineer's experience. Mental comparisons are usually made with similar components and similar conditions, but systematic, formal, procedures are not used. This method, although the least desirable of all, may often be used because of urgent delivery or cost requirements.

The "extrapolation" method of estimating failure rates is simply a more formal application of the educated guess. First, the functions to be performed are analyzed, the conditions to be met are identified, and the components are compared to similar equipment. Then the proposed component is examined to determine how well it performs the needed functions under the specified conditions. Finally, an estimate is then made of its reliability under these conditions, and the estimate compared to the reliability of similar components.

In the "expert opinion" technique, the component elements of a proposed design are identified along with any alternative components and any similar components of similar design. A group of experts are then each asked to rate each component as to its expected reliability in the proposed system. They may be asked to also identify any reliable alternative components or designs, which may have been overlooked, and to rate these also. The designer then averages the ratings and summarizes the comments of the experts. If suggestions are many and/or contradictory, the summaries may be returned to the experts for further consideration. The designer then makes a final summary and studies it to see what modifications can be made to improve the design.

An "actual measurement" can be made by testing prototypes and then calculating the resulting failure rates. The advantage of this method is that the testing conditions can be controlled so that they more closely approximate the required system conditions and thus provide a greater degree of confidence in the data. The problem with this method is the long period of time necessary to obtain the data as well as the great amount of design work needed. This information could not be applied at the very start of a design cycle. If errors were found, extensive redesign efforts would be necessary.

"Failure rate tables" are available to provide required failure data in the cases where design engineers are working with known components. (This is the situation most of the time — most design efforts are with known components where failure data are available.) If the conditions under which the tables were compiled are similar to the desired operating conditions of the system being analyzed, the table values can be used directly. At times, when conditions are different, but not too different, failure rates can be extrapolated from the table data. However, extrapolated data cannot be used with the same degree of confidence as can the direct table data.

12.5 Reliability Prediction

Reliability prediction is a method of determining the probable reliability of a proposed product. This probable reliability is calculated by the use of the appropriate probability distribution (usually the exponential failure law), the previously estimated failure rates, the desired time limit (usually called "mission time" and usually determined by Engineering and/or Management), and the appropriate probability rules. The best way to illustrate the procedure is through an example.

Example 12.1 A design engineer is charged with designing a new radio receiver to be used in space explorations. Most of the components of the proposed design are known with known failure rates. The few that are not have been tested (prototypes were made and used) and failure rates were calculated from the tests. Failure rate and frequency data are summarized in the following table:

Part	Power Supply	rf Stage	Mixer	if Stage	Power Output	Failures per 1000 hr
Resistor	30	40	32	55	45	0.030
Capacitor	15	50	60	80	50	0.035
Transformer	1	5	2	10	1	0.040
Relay	5	3	1	2	3	0.050
Coil	2	20	0	0	0	0.025
Switch	1	0	0	0	0	0.060
Connector	5	4	3	2	2	0.020
Semiconductor A	3	3	0	5	0	0.015
Semiconductor B	4	0	2	0	2	0.018
Semiconductor C	0	3	1	5	0	0.010

If all components (including assemblies) are connected together, functionally, in series, what is the failure rate and 100-hr reliability for each assembly and for the entire receiver?

Solution:

Since the parts are all connected together in series, the failure rates can be added. Failure rate calculations for the power supply are

30(0.030) + 15(0.035) + 1(0.040) + 5(0.050) + 2(0.025) + 1(0.060) + 5(0.020) + 3(0.015) + 4(0.018) + 0(0.010) = 2.042 failures per 1000 hr. Reliability calculations for the power supply (assuming constant failure rates that are exponentially distributed) are $e^{-0.002042(100)}$ = 0.8153. Similar calculations can be made for the rest of the assemblies and for the receiver. The following table summarizes these failure rates and reliabilities.

Assembly	Failures per 1000 hr	Reliability
Power supply	2.042	0.8153
rf stage	3.955	0.6733
Mixer	3.296	0.7192
if Stage	5.115	0.5996
Power output	3.366	0.7142
Receiver	17.774	0.1691

Obviously, a reliability of 16.91% for the ratio receiver would be totally unacceptable.

If, in order to increase the reliability of the receiver, it has been decided to add one backup unit in parallel to each assembly (not each component part), what is the new receiver reliability?

Whenever units are added in parallel (called redundancy), system reliabilities are increased. In this case, the reliability of each assembly is calculated by first multiplying the unreliabilities (1 − reliability) and then subtracting the resultant from 1.00. For the power supply unit these calculations are 1 − (1 − 0.8153) (1 − 0.8153) = 0.9659. The following table summarizes the new reliabilities.

Assembly	Parallel Reliability
Power supply	0.9659
rf stage	0.8933
Mixer	0.9212
if stage	0.8397
Power output	0.9183
Receiver	0.6129

The new receiver reliability is calculated using the multiplicative probability law, by multiplying the five revised assembly reliabilities together.

Adding redundancy by assemblies (called "high-level" redundancy) has increased the reliability of the receiver from 17% to 61%. However, this is almost certainly not enough for the intended, critical, application. Several other things can be done to increase the receiver reliability, as designed. One way is to add another entire receiver (including the redundant assemblies) in parallel with the first. The reliability of this redundant receiver system would be $1 - (1 - 0.6129)(1 - 0.6129) = 0.8502$. Further redundancy could be added to either the assemblies (except for the power supply assembly — it is already too close to 100% for an additional parallel unit to effectively raise its reliability), to the parts (called "low-level" redundancy), or to the entire receiver in order to raise the reliability to an acceptable level; except that the size and weight of the resultant unit would probably be excessive.

An analysis of this nature would highlight the problems to the design engineer even before the design was completed. In this case, more reliable parts are going to have to be found, or designed, before an acceptable, reliable, product can be properly designed. The decision as to which parts should be examined first would be greatly enhanced by a full-fledged FMECA analysis. In Example 12.1, if no more criticality information is available, the parts with the highest total failure rates (the resistor and the capacitor) would be the most critical and would thus be analyzed first.

Another option open to the designer is to add additional parallel units at the individual part level (called "low-level" redundancy). An examination of the failure rate/frequency table reveals that two of the parts, the resistor and the capacitor, contribute enormously to the system failure rate (14.9 of the 17.774 total). If these two parts could be redesigned so that redundancy were integral to the design without adding materially to the weight (a not unreasonable assumption for parts of this nature), the system reliability would be increased to 74% (from the 17% for the original design). Now if redundancy is added to the assemblies as before, the system reliability is increased to 95%. Adding further redundancy at the system level, as before, then raises the receiver reliability to a very respectable 99.75%.

Reliability calculations for the "low-level" redundancy just described will now have to be made at the individual part level rather than at the assembly level as before. (Simple addition of failure rates does not work with this type of problem.) The individual resistor, for instance, has a reliability of 99.7% ($e^{-0.00003(100)} = 0.997$). When multiplied together 202 times, for the 202 total resistors in the entire receiver, the total resistor reliability becomes 54.5%. If the resistor is redesigned to provide redundancy at the individual part level, the new part reliability is increased from 99.7% to 99.999%. This revised part reliability, when multiplied together 202 times, gives a total resistor reliability of 99.8% instead of the 54.5% of the previous design. Similar redesign of the capacitor would raise its reliability from 40.96% to 99.69%.

As can be seen from this example, this type of analysis would not only highlight problems to the design engineer but could also indicate areas where improvement activities should be concentrated and even direct attention to the type of improvement or redesign that could best solve the problem with the least cost. FMECA is indeed a powerful tool for reliability and design analysis.

Note that the above analysis is accomplished using the procedures and formulas of Chapters 6 and 7, which in turn are based on probability and statistical concepts explained in earlier chapters. At this point, management might want to know what mean reliability value the system must achieve in order to be, say, 95% confident that it will meet the minimum reliability criteria of 95%. This would require the techniques explained in Chapter 8. Of course, in order to be meaningful to manufacturing, the required mean life values of each component would have to be calculated.

Let us suppose that management requires a 95% level of confidence that the system can indeed meet the 95% reliability requirements. The 95% reliability would then become the lower confidence limit for the reliability and the mean reliability level must be calculated. In order to determine the mean reliability of the system, the mean life must first be determined. The calculations are as follows:

1. Calculate m_{LC} (R_{LC} = 95%):

$$0.95 = e^{-100/m_{LC}}$$

therefore

$$T = 100 = m_{LC} \ln(1/0.95)$$

and

$$m_{LC} = 1949.573 \text{ hr}$$

2. Calculate $\overline{m}$:

$$\overline{m} = \frac{m_{LC}(\chi_{2f,\,\alpha})}{2f}$$

$$= \frac{1949.573\,(5.991)}{2(1)}$$

$$= 5840 \text{ hr}$$

3. Calculate mean reliability:

$$R_{\overline{m}} = e^{-100/5840} = 0.983$$

The system with all possible redundancies (that is, with all low-level and high-level redundancies combined) is equal to 99.75%, which more than meets the 98.3% mean reliability requirements. As far as design is concerned, this is all the calculations that are needed to support the design requirements. Manufacturing, of course, would need to know the component mean lives required to meet these specifications. These would be determined using inverse procedures similar to those just used above. The new, redundant, resistor (for instance) has a reliability of 0.99999, which calculates to a mean life of 99,995 hr. Manufacturing would also be concerned with the control of the production function, using the procedures outlined in Chapters 10 and 11.

The next design consideration in this problem is the maintenance and maintainability of the system (Chapter 9). The ease of testing and locating faults, easy access to all components, the nature of primary as opposed to secondary maintenance, the difficulty in training competent repair people, etc., must be considered in light of system requirements. It seems obvious that, for the system under discussion, maintenance considerations are critical and repair times must be minimized (maintainability must be maximized). This may not be true, of course, for other types of systems (or may not be so critical).

The final design concerns are safety (Chapter 13) and product liability (Chapter 15). Danger to operating and maintenance personnel must be minimized and even eliminated (if possible). To keep all of these requirements balanced, while designing a safe, efficient, reliable product at least cost, is seldom easy.

Failure Rate Allocation

Sometimes a complex system has to be designed where only overall system goals are known and many of the components are nonstandard. The components must be designed as well as the system. Since only system specifications are given, the component reliabilities must be calculated. Management would like to design only to this required reliability level to save on costs (since reliability and costs are usually directly related—increasing reliability usually increases costs also). Component reliabilities are determined by allocation of failure rates (since failure rate is related to reliability through the applicable probability distribution).

In the allocation of failure rates, system components are assumed to be connected together, functionally, in series; the system is assumed to be static; the failure rates are assumed to be exponentially distributed; and only single, lower, confidence limits are used. Actually, in the initial design, all components are always connected together in series. Adding redundancy increases costs and, so, redundant components are only added after the simple series design proves unreliable (does not meet system reliability specifications). Sometimes components are designed in parallel for other than redundancy reasons, but this design is integral to the component function and, so, the entire parallel unit is considered, for design and reliability purposes, to be one component. Costs are also the reason for designing a static system first, before considering dynamic systems (the elimination of interaction complexity is another reason).

The reason for assuming exponentiality is that these types of calculations are easy to make, and most complex system reliabilities are either exponential or very close to it. Also, the exponential distribution is the only one that requires only one parameter (the mean), which can easily be calculated inversely from specification data. Other distributions require more than one parameter (the normal needs two, for instance, the mean and standard deviation). Since test information is not available at this time, one or more of these parameters would have to be assumed (the standard deviation in the case of the normal). And, finally, only lower confidence limits are used because upper confidence limits, in these cases, are redundant. Why put an upper limit on reliability? The higher the reliability the better.

If any of these assumptions prove to be untenable later, failure rates can then be adjusted to fit. The important thing is that these assumptions allow a relatively easy means of determining, at least initially, the component failure rates; they provide a starting point for design.

The steps in the failure rate allocation procedure will be explained in conjunction with an example. Suppose that a certain system is to be composed of five assemblies; $A1$, $A2$, $A3$, $A4$, and $A5$. The reliability specification is to be 80% at a 90% level of confidence for a 100-hr mission. The $A3$, $A4$, and $A5$ assemblies are standard with known failure rates of 0.0001, 0.0002, and 0.0003 failures per hour, respectively. The relative worths of assemblies $A1$ and $A2$ are 60% and 40%, respectively (the relative worth concept will be explained later). Assembly $A1$ consists of three components — $C1$, $C2$, and $C3$ — with relative worths of 60%, 30%, and 10%, respectively. Find the allocated failure rates and respective reliabilities for the three components. The steps are:

1. Determine the mean failure rate (FR) for the system. This is done by inverse application of the exponential failure law and the confidence limit formula. These formulas are:

$$R_{LC} = \exp[-(T/m_{LC})]$$

$$m_{LC} = 2f\overline{m}/\chi^2_{2f,\alpha}$$

where

R_{LC} = the lower reliability confidence limit

T = the mission time (minimum operational time without a failure)

m_{LC} = the lower confidence limit for the mean life

f = the number of failures in a test (if no test has been made, this number is assumed to be 1 — which is always the case for failure rate allocation problems)

$\overline{m}$ = the average mean life (this is the value that is used as the production specification)

χ^2 = the symbol for the chi-square distribution

α = the level of significance, *or* one minus the level of confidence, *or* the probability that the real universe mean will *not* be above the critical value (the critical value is usually the mission time in these types of calculations)

$\chi^2_{2f,\alpha}$ = the symbol that designates a particular value from the chi-square table (Table A3)

The mean failure rate, then, is calculated as follows (note that some of the formulas are inverse solutions of the preceding general formulas):

(1) $m_{LC} = T/[\ln(1/R_{LC})] = 100/[\ln(1/0.80)] = 448$ hpf

(2) $\chi^2_{2f,\alpha} = \chi^2_{2(1),0.10} = 4.605$ (Table A3)

$\overline{m} = \chi^2_{2f,\alpha}(m_{LC})/2f = 4.605(448)/2(1) = 1{,}032$ hpf

(3) $FR_{-} = 1/\overline{m} = 1/1{,}032 = 0.000969$ failures per hour (note that failure rate is the reciprocal of mean life)

Note: If no confidence level is specified, this is equivalent to a 50% level of confidence. At this level, the mean and the lower confidence level become the same thing and the confidence limit formula is not needed. The failure rate calculation, in this case, is:

$$FR_{-} = [\ln(1/R)]/T = [\ln(1/0.80)]/100 = 0.000223.$$

2. Sum the standard assembly failure rates (where failure rates are known), and subtract from the calculated system failure rate. This determines the amount of failure rate left over to be allocated among the remaining assemblies, those where the failure rates are unknown.

$$FR_{A1} + FR_{A2} = FR_s - (FR_{A3} + FR_{A4} + FR_{A5})$$

$$= 0.000969 - (0.0001 + 0.0002 + 0.0003)$$

$$= 0.000369 \text{ fph}$$

One common problem at this point occurs when the sum of the known assembly, or component, failure rates is greater than the whole (if the sum of FR_{A3}, FR_{A4}, and FR_{A5} were to be greater than 0.000969). Obviously, redundancy must be designed into one or more of the known assemblies before the preceding allocation procedure can continue. Redundancy should first be added to the component (or assembly) with the highest failure rate.

3. List the remaining, nonstandard, assemblies and determine their relative worths. This would be shown as a percentage of the remaining system failure rates. The relative worth figures, therefore, must sum to 1.00 (or 100%). Which assemblies are more important to the system (and, therefore, have a higher relative worth) is a matter of judgment. The term "more important," in this context, means lower failure rate. For instance, an assembly with a high repair cost should have a lower failure rate (higher relative worth) than one that is inexpensive to repair. Costs, safety, repairability, and complexity are among the most important elements of relative worth. In the present problem, this step was already accomplished in the problem statement.

4. Apply the relative worth figures to the remaining system failure rate to determine the remaining, nonstandard, assembly failure rates. Relative worth assumes a type of inverse relationship to failure rate. Since assembly $A1$ has a higher relative worth than $A2$, it must have the lower failure rate. Therefore, 60% of the remaining failure rate must be applied to $A2$, not $A1$ (when there are more than two unknowns, a system of inverse simultaneous equations must be used, as in Step 5):

$$FR_{A1} = 0.40(0.000369) = 0.000148 \text{ fph}$$

$$FR_{A2} = 0.60(0.000369) = 0.000221 \text{ fph}$$

5. The $A1$ failure rate of 0.000148 failures per hour must now be allocated among the three components of assembly $A1$. For three or more unknowns, the allocation calculations become complex, and require solutions to multiple sets of simultaneous equations. These equations are

$$FR_1 + FR_2 + FR_3 + \ldots + FR_n = FR_a$$

$$FR_1 = (RW_n/RW_1)FR_n$$

$$FR_2 = (RW_n/RW_2)FR_n$$

$$\vdots$$

$$FR_{(n-1)} = (RW_n/RW_{(n-1)})\, FR_n$$

where

$$FR \ = \ \text{failure rate}$$

$$RW \ = \ \text{relative worth}$$

$$n \ = \ \text{number of units with unknown failure rates}$$

The solution for the set of three components of $A1$ are

(1) $FR_{C1} + FR_{C2} + FR_{C3} = FR_{A1} = 0.000148$

(2) $FR_{C1} = (RW_{C3}/RW_{C1})FR_{C3} = (0.10/0.60)FR_{C3}$

(3) $FR_{C2} = (RW_{C3}/RW_{C2})FR_{C3} = (0.10/0.30)FR_{C3}$

Substitute the solutions to Eqs. (2) and (3) into Eq. (1) and solve for FR_{C3}:

$$\tfrac{1}{6} FR_{C3} + \tfrac{1}{3} FR_{C3} + FR_{C3} = 0.000148$$

$$\tfrac{9}{6} FR_{C3} = 0.000148$$

$$FR_{C3} = 0.0000987$$

Now substitute the FR_{C3} into Eqs. (2) and (3) to find FR_{C1} and FR_{C2}:

$$FR_{C1} = \tfrac{1}{6} FR_{C3} = \tfrac{1}{6}(0.0000987) = 0.0000164$$

$$FR_{C2} = \tfrac{1}{3} FR_{C3} = \tfrac{1}{3}(0.0000987) = 0.0000329$$

Note that $C1$ is six times more important than $C3$. Because of the inverse relationship of failure rate to reliability (or to importance), this ratio (6 to 1) must be reversed in the formulas. A component that is six times more important than another must have only $\tfrac{1}{6}$th the failure rate. The same logic, of course, applies to $C2$ ($C2$ must have $\tfrac{1}{3}$rd the failure rate of $C3$).

    6. Calculate the individual component reliabilities:

$$R_{C1} = \exp[-(FR_{C1}T)] = \exp[-(0.0000164(100))] = 99.84\%$$

$$R_{C2} = \exp[-(FR_{C2}T)] = \exp[-(0.0000329(100))] = 99.67\%$$

$$R_{C3} = \exp[-(FR_{C3}T)] = \exp[-(0.0000987(100))] = 99.02\%$$

If any of the preceding components cannot be designed to these figures, redundancy will have to be employed. For instance, suppose that component $C3$ can only be designed to a 95% reliability. Two of these components in parallel would give a reliability of $1 - (1 - 0.95)(1 - 0.95) = 99.75\%$ (which more than meets the calculated requirement of 99.02%).

If some of the other allocation assumptions now prove wrong, the preceding figures can serve as a base from which to adjust to reality.

Although the above example contains only three levels (system, assembly, and component), the procedures can be generalized to any number of levels. The computer is obviously well suited to assist in these types of calculations.

Terms and Definitions

1. **Operability.** Ability of a product to perform its intended function efficiently.

2. **Producibility.** Ability of a product to be easily and efficiently manufactured without excessive costs.

3. **Procurability.** Ability to buy quality parts without undue effort, cost, or delivery time.

4. **Expandability.** Ability of the product to be redesigned for multiple goals without excessive effort or cost.

5. **Versatility.** Ability of a product to be used for multiple goals with no redesign efforts.

6. **FMECA (failure modes, effects, and criticality analysis).** A reliability analysis model that directs maximum design efforts to those areas where maximum return can be expected.

7. **Criticality.** A calculated ratio designed to highlight those components that contribute most to a product's probability of failure.

8. **Educated guess.** A method of failure rate estimation using the design engineer's past experiences.

9. **Extrapolation.** A failure rate estimation model using a formal procedure for directing the design engineer's attention to those factors that will best assist in the estimation process. Uses the engineer's past experience along with known information from similar products.

10. **Expert opinion.** A failure rate estimation model using the average of a panel of experts to estimate failure rates. The past experience of the panel is used along with known information from similar products.

11. **Actual measurement.** A failure rate estimation model using actual test information to determine failure rates.

12. **Failure rate tables.** A failure rate estimation model using known failure rates from published data of component performance from other, similar, products.

13. **Reliability prediction.** A method of determining "probable" product reliability (of a proposed product) by reliability calculations using known, estimated, or extrapolated component failure rates and the appropriate probability laws and distributions.

14. **Exponential failure law.** Using the first term of the Poisson probability distribution, or probability of zero failures, using failure rate times mission time as the exponential of the natural logarithm to the base e.

15. **Series systems.** A complex system of many parts connected together so that the entire system fails if one component fails.

16. **Parallel systems.** A complex system of many parts connected together in such a way that the system does not fail when only one

part fails — all parts must fail before the system can fail (also called redundancy).

17. **Redundancy.** The practice of connecting two or more of the same parts together in parallel in order to increase system reliability. Extra units connected in parallel are called redundant or back-up units.

18. **Unreliability (1 − reliability).** The probability that the system will not last its mission time without failure.

19. **High-level redundancy.** The practice of adding back-up units in parallel at the assembly or full system level (high level) in order to increase system reliability.

20. **Low-level redundancy.** The practice of adding back-up units at the component level (low level) in order to increase system reliability. All other things equal, "low-level" redundancy increases system reliability to a greater degree than does "high-level" redundancy. However, "high-level" redundancy facilitates the use of plug-ins (commonly called "black boxes").

Practice Problems

1. Using the following data for a complex system, and assuming an exponential distribution and a 100-hr mission time, determine the assembly and system reliabilities. There are four assemblies and three components with the three components distributed among the assemblies as follows. The failure rate for components $C1$, $C2$, and $C3$ are 0.0001, 0.0002, and 0.0003 failures per hour, respectively.

	$A1$	$A2$	$A3$	$A4$
$C1$	1	2	3	4
$C2$	2	3	4	5
$C3$	3	4	5	6

Answers: 86.9; 81.9; 77.1; 72.6; 40.0.

2. Determine the assembly and system reliabilities for Problem 1 if redundancy is employed at the assembly level (each assembly is composed of two assemblies in parallel).
   Answers: 98.3; 96.7; 94.8; 92.5; 83.3.

3. Determine the assembly and system reliabilities for Problem 1 if redundancy is employed at the component level.
   Answers: 99.7; 99.5; 99.4; 99.2; 97.8.

4. Determine the assembly and system reliabilities for Problem 1 for a 90% confidence level.
   Answers: 72.4; 63.1; 55.0; 47.9; 12.0.

5. Determine the answers to Problem 2 for a 90% level of confidence.
   Answers: 92.4; 86.4; 79.7; 72.8; 46.3

6. Determine the answers to Problem 3 for a 90% level of confidence.
   Answers: 98.2; 97.5; 96.9; 96.2; 89.2.

7. Determine the assembly and system reliabilities for Problem 1 for a 99% confidence level.
   Answers: 52.5; 39.8; 30.2; 22.9; 1.45.

8. Determine the answers to Problem 2 for a 99% level of confidence.
   Answers: 77.4; 63.8; 51.3; 40.6; 10.3.

9. Determine the answers to Problem 3 for a 99% level of confidence.
   Answers: 93.4; 91.0; 88.6; 86.3; 65.0.

10. Determine component and assembly failure rates and reliabilities for the following data:
    a. Mission time = 100 hr.
    b. Reliability = 95%.
    c. Confidence level = 90%.
    d. Five assemblies; assembly $A1$ has a known failure rate of 0.001 fph.
    e. The other four assemblies have relative worths of $A1$ = 40%, $A2$ = 30%, $A3$ = 20%, and $A4$ = 10%.
    f. Assembly $A2$ has three components with relative worths of $C1$ = 40%, $C2$ = 35%, and $C3$ = 25%.
       Answers:

	A1	A2	A3	A4	A5
Failure rate:	0.0001	0.0000147	0.0000196	0.0000295	0.0000589
Reliability:	99.0	99.85	99.88	99.71	99.41

	C1	C2	C3
Failure rate:	0.00000394	0.00000450	0.00000630
Reliability:	99.96	99.96	99.94

11. Solve Problem 10 for a reliability of 99% and a confidence level of 95%.

    Answers: Except for the known assembly, $A1$, the failure rates are all negative and the reliabilities are all greater than 100. Redundancy must be added to $A1$ (two $A1$'s in parallel). With this redundancy, the answers are (failure rates are expressed in failures per 10,000 hr):

	A_1	A_2	A_3	A_4	A_5
Failure rate:	0.009	0.0392	0.0522	0.0784	0.1567
Reliability:	99.99	99.96	99.95	99.92	99.84

	C_1	C_2	C_3
Failure rate:	0.0105	0.0120	0.0167
Reliability:	99.998	99.999	99.983

12. Solve Problem 10 assuming no relative worths (all are equal).
    Answers:

	A_1	A_2	A_3	A_4	A_5
Failure rate:	0.0001	0.0000307	0.0000307	0.0000307	0.0000307
Reliability:	99.00	99.69	99.69	99.69	99.69

	C_1	C_2	C_3
Failure rate:	0.00001033	0.00001033	0.00001033
Reliability:	99.897	99.897	99.897

CHAPTER **13** | **SAFETY**

The purpose of this chapter is to describe the problem of safety in the workplace and its relationship to reliability. Modern safety laws, practices, and principles can have a substantial effect on reliability.

Objectives

1. Understand the history, attitudes, and safety problems that led to modern safety laws.
2. Understand the fundamentals of modern safety legislation — workman's compensation and OSHAct.
3. Know the responsibilities pertaining to employers and to safety managers under the law.
4. Have a knowledge of the methods used by safety personnel in recognizing and controlling hazardous conditions.
5. Understand some of the methods and models used by safety professionals in analyzing and appraising plant safety.
6. Know how safety practices and laws relate to, and affect, reliability design.

13.1 History of Safety

In the early days of the Industrial Revolution the safe operation and use of machines and products received only cursory attention. The development of machines to supplement manual labor was so complex, and took so much of the designer's energy and ingenuity, that safety was seldom considered. In addition, the prevailing attitude was that a man should be able to attend to, and care for, himself (and family). In the predominate agrarian society of the times, this was a reasonable assumption.

However, human beings were no match for fast moving and powerful machines; humans had neither the speed nor the strength to overcome these hazards on their own. Unfortunately, these early attitudes continued to prevail, spurred on by competition. It was frequently not financially feasible for manufacturers to spend the time and the money necessary to equip their machines with safety devices if the competition did not also do so. Since the attitude of so many employers was that costs were more important than human lives, the need for safety laws was apparent. Safety laws would force all employers to provide a safe workplace and so put them all on an equal competitive footing (at least as far as safety was concerned). However, as it turned out, safety laws were not so easy to pass, and, when passed, were not so easy to implement.

Almost from the beginning it was the smaller-sized companies that caused the most problems. Large companies, generally, found that cost and regard for human lives go together. A safeguarded worker, not now having to concern himself or herself so much with safety, was more productive. Smaller companies, however, with smaller profit margins with which to work, could seldom take the chance for safety. Even when consensus standards were agreed upon in an industry, small companies usually broke the agreement sooner or later (owing to cost considerations). Not all large companies were so safety conscious, of course. Witness the statement of a railroad employer who said that he could bury a man cheaper than he could put air brakes on a car.

The responsibility for accident prevention comes from three different, yet interrelated, factors. First is the moral issue that we all have a responsibility to see that others around us are not injured by our actions, even when our actions are negative ones of not providing safety features on a tool or machine. Next is the economic issue where all possible costs are considered in relation to a possible accident (or accidents). Losses incurred as a result of the accident (cost of insurance, insurance reimbursements for losses, and costs of an accident prevention program) are considered in a solely economic analysis. It is with this issue that laws can have such a profound effect by making the accident so costly (especially if a worker is injured) that it far offsets all other costs of accident prevention. Finally, there is the obligation under the common law. The common law says that an employer must provide a safe workplace for the worker, safe tools for the job, competent fellow workers, competent and trained management, and safety rules that are enforced.

With such compelling pressures to provide safety in the workplace, one would think that safety would be almost automatically forthcoming. But, as has been already explained, the problem was extremely complex.

From this interweaving of misunderstanding and economic pressures, five fundamental problems emerged, which finally set the stage for the modern safety laws. They were (1) insufficient compensation, (2) steadily increasing death and injury rate, (3) ineffectivity of early legislation, (4) ineffectivity of "consensus" standards, and (5) rising employer/employee antagonisms.

Four factors combined to prevent most injured workers from gaining adequate compensation for their injuries. First, the burden of proof was on the injured workers, and the technical details and cost of developing this proof were often beyond their capacity. Second, the proof often required the testimony of fellow workers, who were reluctant to testify for fear of losing their jobs. Third, the time in court was so extensive (from 6 months to 6 years) that most injured workers could not afford the long wait without jobs or income. Finally, there was the problem of the biased courts — biased in favor of the employer. The court decisions became almost completely concerned with the technical concept of exemptability (how employers could exempt themselves under the law) rather than with the employer's responsibility under the common law (provide safe workplace, tools, etc.). This exemptability concept led to three exemptions that employers were able to use to win their cases (to absolve their responsibilities under the common law):

1. The concept of "contributory negligence" meant that any action by the plaintiff (injured worker) that helped to cause the injury, or extend the severity of the injury, was cause enough to dismiss the case. The natural and normal proclivity of human beings to forget, do the wrong thing at the wrong time, fail to notice a hazard, etc., was simply ignored by the courts. If injured workers wanted relief in the courts, they often had to almost prove themselves to be infallible. On the other hand, the courts tended to ignore the "contributory negligence" of the employer who provided the workers with unsafe working conditions, tools, etc.

2. Negligence of a fellow employee, if the negligence helped to cause the accident, could block the injured worker from collecting from the employer. The only recourse was to sue the negligent fellow employee, who never had enough money to make this action worthwhile. This was the case even though the employer's responsibility to provide competent fellow workers was clear under the common law.

3. When a worker accepted a job, there was an "assumption of risk" concept applied by the courts. The worker was supposed to examine the job and, if the risk was too great, refuse to do it. Of course, if any worker really tried to do this, he or she would immediately be fired. The inequities of power involved (the employer with power

224 _____ SAFETY

over the continuation of the employee's income versus the employee would could only quit and try to find another job, which were frequently unavailable, especially in areas of only one employer) was, for the most part, ignored by the courts (the inequities of power is a legitimate concept of law). The "assumption of risk" concept became so bizarre that an employee who was injured as a result of driving a defective truck was unable to collect damages because the court found the employee had a better opportunity to check the truck for defects than did the employer and therefore was "assuming the risks" inherent in the defects.

The steadily growing death and injury rate quickly became apparent to everyone. Although the national safety council estimates accident costs at about 50 billion dollars per year, most professionals believe the real cost to be as much as 10 times greater. Many more accidents occur than are reported, owing to many factors (most minor and disabling injuries are not reported; toxic injuries frequently take long periods of time to become apparent; employers transfer injured workers to less strenuous activities to reduce their lost-time reports; many industries are not required to report injuries; etc.). Even in the face of this growing injury and death rate, employers generally made little effort to overcome hazardous conditions. As has already been explained, human life was all too often equated with costs, with costs coming first.

Because of the bias of the courts toward the employer and the exemptability attitude, early legislation was largely ineffective. When effective laws were passed, they were found to be unconstitutional. In order to overcome this, legislatures resorted to consensus-type laws that had little real power to alter the attitudes or the rising injury rate. This is the reason, in fact, why many safety laws today (especially those based on the early workmen's compensation laws) are not as effective as they might be.

Although the picture thus far appears to put much, if not all, the onus for the poor safety record on employers in general, many industries did, in fact, move to correct the problem by agreeing on safety standards among themselves. However, since little or no enforcement was possible, these "consensus" standards were frequently violated owing to cost considerations. Also, the standards agreed upon were never complete enough for real safety effectiveness. In general, the larger companies tended to keep these standards (larger profit margins, perhaps), while smaller companies did not.

Finally, the antagonisms between employers and employees were increasing at an alarming rate owing, to a great extent, to the increasingly poor safety records. Unions repeatedly made safety their number one demand. Social scientists were quite concerned that these feelings could erupt into class wars that might drastically affect our American way of life. Social literature abounded with these and other inequities associated with the Industrial Revolution. The time was right for major new steps for overcoming this safety problem.

13.2 Modern Safety Legislation

Early safety legislation, especially as it applied to compensation for injury, was first found to be mostly unconstitutional. New York, which usually set the standard for the nation, was the first to find a workman's compensation law unconstitutional. Other states then tried to pass modified laws that eliminated the objections of the New York law. Unfortunately, these laws did not have much power. They depended too much on consensus-type agreements. Washington state, however, refused to go along with the New York decision. They reasoned that the time had come for a strong workman's compensation law and that the New York decision was hasty and ill advised. They, therefore, passed a strong workman's compensation law that was upheld by the state supreme court. In 1917, five years after this decision, the U.S. Supreme Court also upheld the law. In the meantime, several states had passed the weaker consensus-type laws. The rest of the states, District of Columbia, and Puerto Rico then all passed their own laws patterned after the strong Washington state law. However, these laws are all different with different court interpretations from state to state making for a chaotic condition of laws and interpretations that are very difficult to keep straight, especially for interstate firms.

The main provisions of the workman's compensation laws are as follows:

1. Liability without fault.
2. Injured workers, and their families, are assured of some benefits, although never enough to offset loss of income.
3. Employers are not allowed to use the three common law defenses (exemptions).
4. Injured workers receive payment for their medical bills.
5. Some benefits are given for permanent injury, although most of these benefits are below subsistence level.
6. The family receives some death benefits.

Among the 54 different workman's compensation laws across the nation, there are four main types:

1. Most are compulsory for both employer and employee.
2. A few allow employers to choose whether or not they will be under the law. (These are the consensus-type laws mentioned previously.)
3. A few states allow some types of businesses to elect, or choose, to

be covered under the law, while others, considered more hazardous, are compulsory. (The changing definition of the term "hazardous" is slowly expanding the compulsory coverages.)

4. A very few are compulsory for the employer but effective for the employee.

It is estimated that approximately 80% of all workers in the United States are covered by workman's compensation laws. Some of the occupations not usually covered (some are in some states) include agricultural, domestic, charitable institutions, and small firms of 2–10 workers (different in different states). Eleven states cover only "hazardous" occupations. However, the definition of "hazardous" has been enlarged to cover most occupations.

In order to gain benefits under the workman's compensation laws, the accident must occur during the course of employment. What the words "course of employment" really means can be, and is, variously interpreted. In one state a man traveling for his company was denied benefits when he contracted food poisoning in a restaurant, while in another state another man received benefits for a similar illness. Another problem was the definition of an accident. The original definition was "a sudden, adverse, unexpected event." However, when psychological and stress-related diseases were admitted by the courts, it became obvious that the definition of "accident" must be revised. Today the definition is the same as before, but with the word "sudden" deleted.

Workman's compensation as a giant step in the alleviation of the problems caused by unsafe working conditions. It now made safety prevention programs more economically feasible as well as ensuring some alleviation of the financial burden for the injured worker and his or her family. However, several problems remained:

1. None of the laws included any provision for maintaining the injured worker's wages at preinjury levels.
2. Workman's compensation payments are woefully inadequate, most of them being well below the poverty level.
3. In general, the large companies found it advantageous to institute safety programs and to reduce unsafe conditions. However, the small companies, for the most part, still found it necessary (owing to cost considerations) to take chances under the law and hope that no major injury would occur.
4. The 54 different laws (50 states, District of Columbia, Puerto Rico, and two federal laws) and the many conflicting court decisions and interpretations make for chaos in the law.
5. Workman's compensation laws concentrate too much on compensation and not enough on prevention. Something else was needed to force accident prevention programs and support safety consciousness. That something was OSHA.

In 1970 the Occupational Safety and Health Act (called OSHAct) was passed. It covers about 5,000,000 plus businesses, from the very small to the very large. Only federal workers and the armed services were exempted, although they have since been included by presidential directive. The objective of the act, as stated in the law, was to ensure a "safe and healthful working condition for every man and woman"; to "preserve our human resources." This objective is to be achieved by five different means:

1. OSHA administration is to encourage all employers to provide safe working conditions, encourage employees to work safely, and encourage states to enact laws and set up procedures to ensure safe and healthful workplaces.
2. Mandatory standards were set and enforcement procedures were instituted.
3. A research branch was organized to provide accurate and comprehensive safety information.
4. Procedures were set up to gather, analyze, and disseminate safety, hazard, and accident information.
5. Procedures were instituted to provide safety training programs.

The responsibility to administer and enforce the new regulations was given to the federal government or to those states that could prove that their laws were "equal or better." Employers were required to (1) ensure safe workplaces (a reiteration of the common law), (2) comply with OSHAct standards, (3) keep records of deaths, injuries, illnesses, and any exposure to hazards and to make periodic reports on all these conditions, (4) keep employees fully informed about all hazardous conditions in the plant, and (5) provide safety training for employees (general training for all and specific training for those in contact with specific hazards). Employees were given the responsibility to comply with all safety rules and to report all violations.

The law provided for a series of safety inspections to be made by OSHA personnel. Originally, the inspections were to be made without prior notice, but the 1978 revision to the law changed this. Now employers must be given a 24-hr notice prior to the scheduled inspection. Periodic inspections are to be made at random until all firms are inspected. Also, any employee can request that a special inspection be made. The request will be analyzed and a determination made as to whether or not to honor the request. The requester will always receive an answer and his or her anonymity will be maintained at all times. Inspections are also made at a death or injury or for any serious, imminent, hazard.

When a violation is found during an inspection, a citation is written and given to the employer. The citation must include a clear, written, explanation of the violation, a reasonable time limit in which to correct it, and a statement of the fine imposed, if applicable. If the danger is imminent, an injunction can be obtained from the courts to close down the operation. Serious violations carry a mandatory penalty, while the amount of the fine for nonserious violations is at the discretion of the inspector (within limits defined in the law). De minimus citations (minor violations that in themselves can cause no danger) carry no penalty. An employer can be fined for up to $1000 per day for each day the violation goes uncorrected. A maximum of $10,000 or six months in jail or both can be assessed for "willful" violations. If a death occurs as a result of a "willful" violation, the penalty is doubled. (Of course, "willful" violations may also be tried under the criminal law.)

Employers have 15 days to reply to a citation or to correct the violation. They may ask for an extension and, if so, they usually receive it (unless in involves imminent danger). In general, the application of the law under OSHA has been quite fair to employers and in most cases quite generous in granting extensions.

OSHAct has had a decided impact since its inception. Accidents have been reduced, especially in the small companies. Many, if not most, of the large companies had already moved to correct their safety problems. Probably the most far reaching effect has been the change in attitude about what causes an accident. Governments, and finally the courts, have come to believe in eliminating accidents by eliminating bad designs. No matter how badly the worker errs (the new logic goes), an accident can happen only if design characteristics are present to allow it to happen. Perhaps if designers can be trained to the new safety attitude, all possible accidents can eventually be prevented by proper design of equipment and products. That this can affect reliability, perhaps drastically, should be obvious.

13.3 Safety Management

In order for a safety program to be effective, it must have whole-hearted top management support. There are two ways that management can give this type of support: by investing the program with the authority it needs and by supporting it with management's influence. Unless the safety program has a clear and explicit authority, it can achieve very little. But this authority must be more than mere lip service or occasional memos. Top managers must actively support the safety engineer's efforts and lead in coordinating and integrating the safety activities of all departments. Management must also show a safety attitude, apparent to all, by continuing to sustain all accident prevention and safety activities.

OSHAct clearly places the responsibility for safety on top management. The employer is responsible for safety, under the act, and cannot delegate this responsibility away. Even if orders are disobeyed, the top-level manager is considered at fault. He or she should have made sure the orders were obeyed. Under the law, both money fines and jail sentences can be levied for flagrant offenses. The money fine can, of course, be paid by the company, but a company cannot serve a jail sentence. Under OSHAct that sentence will be served by the highest-level manager holding clear responsibility for the hazard.

"Careless workers cause most accidents" has been almost a truism in the past; one that has seemed so apparent that it has been accepted by company management, by society and by safety professionals. It has repeatedly been used as a slogan in attempts to reduce accidents. Although various studies and reports on accidents have supported this idea in the past, careful analysis has shown the bias of these reports. Poor design, poor methods, and human nature combining to cause accidents apparently did not occur to the authors of these reports. They were so convinced that the accident had to be caused by a careless human act that they could look for one until they found it (fairly easy to do since human carelessness, unfortunately, abounds), after which the investigation would end with another accident labeled "careless worker." When accident investigators began to extend their analyses to the real cause, it was found that careless workers actually cause only about one-fourth of all accidents. Even these accidents probably have subsidiary causes that made them possible. (Perhaps the accidents might not have happened if careless human acts were the *only* underlying cause.) The modern thinking, then, is that most accidents are caused by careless design where the human element is not considered. Injury is impossible without the presence of a hazard. The problem in the past has been that this attitude on the part of management (that careless workers are the underlying cause of most accidents) has blinded them to the real issue. Management concentrated so much on training workers to be careful, and assessing blame when the workers were not, that the real cause (that of poor design) was not addressed, or even suspected.

Another attitude on the part of management that has adversely affected safety programs is that "safety requirements adversely affect production." This may be true to some extent, but studies have shown that when hazards are removed from workers' jobs, and they have thus been reassured about their safety, workers are more productive. In most cases the greater productivity of the worker should more than offset any added costs due to safety requirements. Unfortunately, first-level managers, who must ultimately see that safety requirements are implemented, are so beset by production schedules that safety usually comes last with them. For first-level managers (foremen, etc.) to become thoroughly trained in safety principles and practices is an absolute necessity.

The costs of safety are largely determined by management's actions. The least expensive action, and the least effective in preventing accidents, is to set procedural rules. Rules are necessary, but when they are relied upon totally, the accident rate is usually not much reduced (as has been the case for so many years). Another problem with reliance upon rules alone is that the courts today are not accepting this as adequate. Some kind of add-on safeguard — the next management course of action — would be the minimal acceptable action in the courts today. The third course of action is to replace the hazardous equipment with equipment that is specially designed for safety. This third alternative is by far the most costly, but also provides, by far, the best accident protection.

Basically, OSHAct provides that hazardous equipment be replaced or people protected at any cost. This has been a tremendous boon to safety personnel who now do not have to justify economically a new piece of equipment they have recommended for safety purposes. They only have to prove the old one hazardous. It is of interest here to note that OSHAct does not provide for penalties to equipment manufacturers but to users of the unsafe equipment. Equipment that is safe in one environment may not be in another. Under the law, it is up to the user (the manufacturer who uses the machine for production) to make sure it is safe.

The following list of management responsibilities is not intended to be all inclusive, but is presented, instead, as a summary of the most important management safety duties. Management may, and does, delegate these duties, but it cannot abrogate its responsibilities under the law:

1. Safety policies must be formulated, put in writing and disseminated to all employees. All employees must understand the safety policies and must understand that management expects them to be obeyed.
2. Safety responsibilities should be assigned clearly to the various departments.
3. A safety department or element (depending on the size) should be established.
4. Establish a safety training program.
5. Investigate all hazards and accidents.
6. Continuously monitor operations, especially hazardous ones, for safety.
7. Keep records of hazards, accidents, and exposure to hazards.
8. Maintain an attitude of safety and communicate this attitude to all employees.
9. Correlate safety activities among departments.
10. Set and approve safety budgets, and provide the needed money.

As has already been stated, management has the responsibility for safety under the law. However, they can, and should, delegate the authority and activities of safety to others. One of the most important of these is the safety engineer. Safety engineers should be generalists with a wide range of knowledge in the technical, legal, and administrative aspects of safety. When details in a particular technical discipline are needed, they call in the experts in that area to assist them. The following partial list of a safety engineer's activities and responsibilities will serve to illustrate the scope of a well-run safety program. Note that many of these responsibilities are almost duplicates of those given to the manager. However, the big difference is that the manager has the responsibility in these areas, while the safety engineer's job is to assist the manager in discharging these safety duties:

1. An up-to-date knowledge of all safety laws, including workman's compensation and OSHAct, and the latest court decisions that might affect their particular industry. Advise management on the laws and on the possible effects these laws might have on their firm.
2. Keep records and make the required workman's compensation and OSHA reporting.
3. Assist in developing safety policies.
4. Monitor activities for safety.
5. Liaison with other departments on their safety efforts.
6. Develop a good personal relationship with all employees and management, especially first-line management, and encourage them in their safety efforts.
7. Approve the safety aspects of plant, equipment, methods, installation, design, etc.
8. Make sure appropriate warning signs are posted at all hazardous areas, exits, etc.
9. Assist top management in developing a safety training program.
10. Investigate all accidents for management.

Record keeping and reporting, a special problem for the safety engineer, deserves further discussion. A file listing all activities should be kept by the safety engineer. This file, if kept properly, will serve to organize the engineer's efforts by making all facts readily available to support any contention and to document all safety, hazard correction, and accident prevention activities. Presentations by safety personnel are more convincing when backed up by a file of facts rather than just the engineer's memory, and management is more likely to give a suggestion more serious consideration when it is a "matter of record." Although at present only top-level managers are being held responsible for hazardous conditions, it is not impossible in the future that an engineer could be held criminally responsible along with the manager.

Engineers, then, may be called upon to prove they at least tried. The safety file would be their proof. The safety file could and should also be used to support and document top management's safety efforts as well as the safety engineer's. Finally, this file provides the information needed for the required OSHA reports.

13.4 Hazard Recognition and Control

A hazard is defined as any condition or action that could lead to an injury or damage. Note that the hazard is the condition (slippery floor, open switch, etc.) and not the possibility of an injury (fall, electrical shock, etc.). Hazards are classified as primary, initiatory, or contributory. Primary hazards are those, usually very few, that are the direct and immediate cause of the accident (slippery floor and smooth shoes). An initiating hazard is that primary hazard that began the action that caused the injury (slippery floor). Contributory hazards usually will not cause an accident of themselves, but will help cause an accident, or increase the severity of one, when combined with one or more primary hazards (not seeing the slippery floor).

There are two major methods of recognizing and determining the presence of a hazard. The first is by analysis and comparison. This usually requires the use of a model such as FMEA (failure modes and effects analysis), fault tree analysis, contingency analysis, and checklists. Checklists make use of the experience of experts, which are reduced to lists of questions to be answered as the area is inspected. Possible hazards are identified, which are then further investigated to determine if such a hazard actually exists and what to do to control it. Suggestion programs make use of the experience of plant workers. After all, they are the ones that really know what is going on.

An excellent formal model, that takes advantage of the experiences of the plant personnel, is the "critical incident technique." Personnel are asked questions about near accidents over a period of time, say, the previous month, as well as errors, mistakes, difficulties, and possible hazard conditions. This, after many interviews, gives a large population of possible causes that, when analyzed, show similarities that indicate areas for further study.

The second method of determining hazards is by the use of the experience and training of the safety engineer. There are ten major types of hazards, each requiring detailed knowledge peculiar to their own nature.

1. *Falls, falling objects, and impacts.* Both size and acceleration affect the severity of injuries caused by falling objects. A small object dropped a few feet might cause only a minor injury, while the same object dropped 100 ft or more, or thrown from a fast spinning machine, could cause extensive damage. The ability of various parts of the body

to resist different types of impacts must also be part of a safety engineer's training.

2. *Mechanical injuries.* Cutting, tearing, shearing, crushing, breaking, and straining injuries are the types of mechanical injuries. Except for one, these types of injuries are mostly prevented by various types of machine guards and safety devices. Straining is the only type of injury that is considered to be almost totally caused by the injured person.

3. *Heat and cold.* Actually heat and cold are not the real conditions but are only the body's perception of molecular activity. High molecular activity is perceived as heat, while low molecular activity is perceived as cold. Heat injuries are burns, heat exhaustion, heat stroke (or sun stroke), tired and listless feeling (leading to loss of production), and various types of psychological reactions. Cold injuries are chilblains, frostbite, freezing, and burns (from cryogenic materials near absolute zero). Excess heat also increases materials' susceptibility to fires and therefore increases fire hazards. A special type of eye injury, fortunately not permanent, is the actinic burn caused by strong flashes of light, such as caused by arc welding operations. Control of heat- and cold-type injuries would include limiting exposure, insulating the process, insulating the worker, and first-aid training for all (to help limit the severity of the injury once it happens). It is important to realize that well-intentioned but ignorant first-aid assistance can frequently be worse than none at all (moving a person with a broken back, cooling heat exhaustion, etc.).

4. *Pressure hazards.* These hazards include pressure-vessel accidents, discharges from safety valves, whipping hoses and lines (caused mostly by breakage), waterhammer (sudden increases in pressure), leaks, vacuums (negative pressure), decompression illnesses, and compressed-gas-cylinder accidents. It is important that pressure vessels be properly designed and then used only for the conditions intended by the design. Safety valves to relieve excess pressure, built-in safety margins (extra strength in case pressure does exceed design specifications), and fail-safe devices (prevent coolant water from being shut off inadvertently, for instance) are some of the means of designing safety into pressure vessels. Safety valves should be placed so that they discharge to the rear, or to where no one is working, to prevent burns (in heated pressure vessels) or even punctures or cuts (in high-pressure vessels). The best way to minimize line leaks is to minimize or eliminate hose connections. Decompression illness is a special problem that needs special training by people who are working in high-pressure environments. Safety measures include proper breathing-gas mixtures, trained personnel, and decompression chambers. Compressed gas cylinders can cause great damage if the valves are in-

advertently broken (they become very dangerous missiles). Care in the design of the valves, in storage of the cylinders, and in the training of personnel can reduce these types of accidents. Another compressed-gas problem is the inadvertent mixture of two volatile gases (oxygen and hydrogen or acetylene, for instance). For this reason, compressed gas vessels should be stored separately and in fairly small lots (to minimize problems if they do occur).

5. *Electrical hazards.* Electrical shock, equipment failures, fire ignition, overheating, explosion, and inadvertent activation are the main electrical hazards. Electrical shock, the sudden and inadvertent stimulation of the nervous system by an electric current, can be deadly, depending on many factors: the amount of the current (the most important factor), the path of the current, the electrical resistance of the body part, dampness, etc. Shocks are caused by bare wires, dampness, insulation failure, and inadvertent activation of a circuit. Shocks can be prevented by frequent inspection of electrical equipment for damaged insulation, bare wires, and dampness; by proper design of the equipment to minimize the effects of these hazards (such as three-wire connectors); and by locking a circuit so that it cannot be activated during repairs. Equipment failure can be especially dangerous with some types of equipment, for instance, heat-treating furnaces with cracked gas atmospheres. If the current fails, the operator has only a few minutes to burn off the gas before the temperature drops to the explosive range. [Cracked gas atmospheres are usually explosive *below* 800°F (427°C)]. Explosions can also be caused by overheating in transformers.

6. *Fires.* The control of fires, with prevention first and suppression second, is one of the most important jobs for the safety engineer. In theory, at least, fires are easy to control. Three conditions must exist simultaneously for a fire to start or continue; a fuel, an oxidizer (oxygen, flourine, chlorine, and many others), and an ignition source. Suppress any one of these and a fire is impossible. In actual practice, however, it is a complex and dynamic problem. So many things are happening in the modern industrial plant that conditions can suddenly occur that will support a fire. For instance, it had long been thought that carbon tetrachloride was not combustible. Now, however, it is known that under certain conditions it will burn violently. Trained personnel in all areas of the plant who understand the problem and constant inspections are necessary to control the fire problem. Some places that are especially susceptible to fires are liquid fuels, combustible liquids, sprays and mists (all materials are made more combustible when in the form of fine spray or mist), and some solids. Some materials are not combustible of themselves but can be made so in the presence of a catalyst or when mixed with another material. The rapid increase

in the use of different types of materials in modern society emphasizes the need for the safety engineer, and the design engineer, to keep current on their effects and interrelationships.

7. *Explosives*. Probably the one most important safety activity in relation to explosive material is to keep it widely separated in small amounts so that if an explosion does occur, the damage will be minimized. The problem is that modern industrial plants are using so many new materials in so many new ways that materials that are not normally explosive in one condition might be so in another. Also, many normal and regular processes can be potentially explosive if not carefully controlled. Two examples already mentioned are electrical transformers and fired pressure vessels (boilers). Explosive inhibitors can sometimes be mixed with a material to suppress its tendency to explode. Almost any material is explosive when in the form of a fine dust. In this case, good housekeeping is an absolute necessity. Australian grain elevator operators have been highly successful in controlling grain explosions during a time when such explosions were common in the United States. They did it, primarily, with good housekeeping procedures.

8. *Toxic materials*. With the increasing use of exotic materials, more toxic materials have entered the environment. The great danger of so many of these new materials is that their toxicity is often undetectable and their damaging effects often take a long time, in some cases years, to show up. It is important for the safety engineer, and the design engineer, to understand these effects on the physiology of the body, how the materials are absorbed into the system, and the precautions that need to be taken to prevent damage to personnel. Toxic materials are defined as those materials where only a very small amount of the material is needed to cause injury to the average adult. The effects of a particular material are usually so complex a problem, so many variables involved, that injury dosages are uncertain at best. One important variable is the entry method — how the material enters the body. Entry can come through the skin, gastrointestinal tract, or respiratory system. The injuries from each method depend on the nature of the material. Some materials are readily absorbed through the skin, while others are not, and some materials are diluted by the gastrointestinal tract, while others react with gastrointestinal chemicals causing toxic by-products. Some materials may cause minor injuries at first but systemic, organic, injuries after a period of time. Respiration entry is considered the most injurious because it enters into the physiological systems immediately. There are eight basic types of toxic agents, depending on their physiological effects:

1. Asphyxiants interfere with the body's ability to use oxygen. Some

interfere with the blood's ability to transport oxygen, some with the cells' ability to absorb oxygen, and some with the blood circulation, while others have a combination of effects.

2. Irritants cause inflammation of the tissues, which leads to edema (retention of liquids). If this happens in critical areas or organs, such as the lungs (pulmonary edema), injury and death can result.
3. Systemic poisons cause injuries to internal organs and are frequently reactive with body chemistry.
4. Anesthesia causes loss of sensation, depressing the nervous system, and can cause respiratory failure.
5. Neurotics affect the central nervous system and can be either depressants, or stimulants, depending on the material. Some neurotics can act as stimulants first and then depress, and some vice versa.
6. Hypnotics act to induce a sleep.
7. Carcinogens are usually quite long acting and cumulative, eventually causing cancer.
8. Corrosives react with and harm the skin. Injury can vary from a bad burn for a strong acid or alkali to dermatitis for weak corrosives.

9. *Radiation.* There are two types of radiation: ionizing and non-ionizing. Ionizing radiation causes damage by dislodging an electron in a cellular component that then ionizes (changes the electrical charge) of that particular cell, which in turn will change the function of that cell. When enough of the cells are ionized, the changed function of these cells can cause harm to the body, especially if they happen to be in a critical organ. There are four types of ionizing radiation — alpha particles, beta particles, x-rays, and gamma rays. The large alpha particle (helium nuclei), unless localized and intense, pose little danger to the body since they cannot penetrate the skin. Beta particles (free electrons) can penetrate the skin somewhat and thus are greater hazards to the body. Probably an even greater hazard of the beta particles is the secondary radiation they can cause when striking shielding and other materials (such as oxygen in the air). X-rays (generated with electricity about 15,000 V) and gamma rays (from a radioactive material) are very hazardous, with great penetrating power, even at long distances. They must be heavily shielded and carefully controlled at all times.

With ionizing radiation, the effects are cumulative; therefore, exposure meters and careful records are necessary. Of course, the greater the intensity of the radiation source, the shorter the time needed to cause permanent damage. With very high radiation sources, the critical time can be measured in microseconds. Sources of ionizing radiation, in most modern industral settings, are (1) nondestructive testing (x-rays and gamma rays); (2) medical facilities; (3) chemical and radiation laboratories; (4) neutralization of static electricity (beta particles); (5) thorium from welding rods and magnesium alloys; (6) high-power, high-voltage operations such as electron beam welding and cutting (mostly from secondary radiation from the air); and (7) radioisotope contamination.

Unfortunately, the danger from nonionizing radiation has only recently been recognized. There is great controversy over what constitutes a dangerous dose (10–10,000 μW) and very few standards exist as yet (OSHA has very few). Nonionizing radiation consists of ultraviolet light, variable light, infrared light, and microwaves. Ultraviolet causes burns such as sunburns and photochemical reactions such as conjunctivitis (caused by intense flashes of light from welding operations, reflections of the sun from snow, lasers, plasma arc, etc.). The light must be rather intense and the exposure fairly long to cause any damage, which is confined to eye damage only. Infrared radiation converts to heat causing thermal-type damage (burns). Microwaves affect the kinetic energy of absorbing tissues causing heat (by inducing a current in the material) as in microwave ovens. It is this type of nonionizing radiation that has caused all the controversy. Since there is so much of it in modern life, it is almost impossible to get away from it. If it is true that only a small amount of microwave activity (over 10 μW) can cause some problems (birth defects, sterility, cataracts, cancer, and behavioral changes), a lot of shielding and corrective action will be needed in the future to overcome the problem.

10. *Noise.* Noise is another hazard that has only recently been appreciated. Its affects are cumulative and tend to take years to become apparent. In fact, until recently, loss of hearing has been accepted as a risk of growing older. But recent investigation has shown that most hearing loss in the older years is related to noise levels throughout life (which cause cell deterioration in the inner ear) rather than the so-called "normal" deterioration of old age (which would show up as a change in the small bones of the middle ear). Experiments have been made that suggest that background noise in modern factories is much too high, well above the OSHA standards of 90 dB. In fact, these experiments suggest that the OSHA standards themselves are too lenient. (Apparently, noise damage begins at 80 rather than 90 dB.) Two types of controls are used to protect people from the noise damage: administrative action and engineering controls. Administrative action consists of rules of conduct to protect employees from noise and requiring then to wear ear plugs and ear muffs (or both in extremely noisy environments). Engineering controls include eliminating the sound generators by using quieter motors, preventing vibration (solid foundations), using dampening equipment, isolating the noise sources, and redesign of machines for quieter operation.

Hazard Control

One of the most important jobs of the safety engineer is to keep potential hazards under control so that they do not erupt into injury, damage, or death. The steps the engineer would use to accomplish this task are as follows:

1. List all possible hazards.
2. Estimate the type of possible accidents and the amount of possible damage for each.
3. List the primary and contributory hazards likely to be involved in each accident.
4. Provide safeguards to prevent each type of accident.
5. Establish procedures for rapid and effective reaction to accidents when they do occur, including alarm systems, transportation services, proper emergency equipment at the proper location, trained personnel, safety zones, evacuation routes, and shelters.
6. Designate fire and emergency exits; ensure that they are clearly marked (OSHA law); and establish procedures to ensure that they are never blocked or locked.
7. Enforce good housekeeping (the first step in a safe and productive operation).
8. Provide for first-aid training programs so that good intentioned but ignorant assistance (often worse than no assistance at all) in an injury will be eliminated or, at least, minimized.
9. Promote failure rate minimization as part of the design of products, machines, and procedures.
10. Provide for the monitoring of all hazards by providing warning devices and by periodic checkups and inspections by safety personnel.
11. Ensure that necessary safety rules are instituted and enforced.
12. Ensure that safety training is provided as needed.
13. Maintain an awareness of safety by posting hazardous areas, billboard displays, posters, safety displays, competitions, persuasion, etc.

13.5 Appraising Plant Safety

The purpose of appraising the plant for safety is preventive in nature; that is, to prevent accidents rather than just correct the problem after an accident has occurred. It is in the new plant, or one just being proposed, where safety engineers can have the greatest impact. They should review proposals and plans for safety, note where safety problems are likely to be found, suggest changes to correct the problems, and follow through to see that the safety problems have been corrected. In this way the engineer can see that safety is designed into proposed buildings, equipment, and methods. It is much better to eliminate hazards at the design stage than to expect to control them later through safeguards and safety rules.

Existing plants and equipment, on the other hand, may not have been designed with safety engineering assistance. (Even today, safety engineers are seldom called in to assist in design of plant, equipment, and methods.) Therefore, the purpose of appraising existing plants is to locate existing hazards and to suggest ways of eliminating or controlling them. Unfortunately, cost considerations seldom allow the elimination of a hazard once it is part of the system. All too often, control of these hazards must depend on safety rules rather than designing or redesigning the hazard out of the system.

The most effective safety appraisal methods, at present, are qualitative rather then quantitative in nature and thus depend, to a great extent, on the expertise of the appraiser (safety engineer). Quantitative models are available but they depend too much, from necessity, on statistical procedures that must have large groups of data in order for the results to be meaningful. The problem is that a large amount of data about a particular type of accident in a particular plant may take years to accumulate during which changes in procedures and equipment would tend to invalidate the statistical results. (Changes would be certain if the accidents were at all frequent.) There are, however, several models that show promise.

Checklists

A review of the plant using checklists to guide the evaluation is, at present, one of the most effective means of appraising plant safety. It, of course, depends on the knowledge and experience of the engineer doing the appraisal. However, the model formalizes this procedure to ensure that nothing is forgotten. The first step is to review the functions to be performed and the environment of each function to see if any of them can possibly have an adverse effect on personnel, equipment, etc. The hazards thus identified should be listed on a sheet of paper along with the cause, effect, probability of an accident, type of possible accident, and corrective or preventive measures that must be applied. Checklists are used at all stages of the analysis as an adjunct to the engineer's expertise and to ensure that nothing has been forgotten.

Failure Modes and Effect Analysis (FMEA)

FMEA is used to analyze equipment to determine what the effects of individual components might be on the entire assembly or equipment. First, major assemblies of the equipment are listed, after which each assembly is broken down into its component elements. Each component is then studied to see how it could fail, what could cause each type of failure, and the effect of this failure on other components, subassemblies, and the entire product. Failure rates (expected failures per period of time) can then be estimated (if failure rate information is not available, the part can be compared to similar parts where such information can be acquired) and calculations can be made to determine the overall probability that the product will operate for a certain period of time, or the probable operating time between failures. Unfortunately, this method is limited to determining causes and effects of equipment malfunctions.

Fault-Tree Analysis

Fault-tree analysis uses the concepts of Boolean algebra to determine the probability of a predetermined event occurring. Each subevent that can contribute to the event (called a "top event") is chosen in such a way that, in relation to the top event, it can assume only one of two possible conditions — either/or, plus/minus, 0/1, on/off, etc. (a dichotomous relationship). Each of these events is then further subdivided in exactly the same way until no further subdivision can take place. When these are charted, the chart becomes a fault tree (Fig. 13.1). The chart looks something like an upside down tree with the top event forming the trunk and the subevents forming the limbs and branches. Analyzing "down" a tree progresses from effect to cause (or causes), while analyzing "up" a tree progresses from cause to effect (or effects).

The "top event" is chosen by analysis, using the checklist method, from a list of potentially hazardous conditions already known or strongly suspected or from specification requirements. Of course, there can be more than one "top event" (and usually are), and there will be as many fault trees as there are "top events." The events are not limited

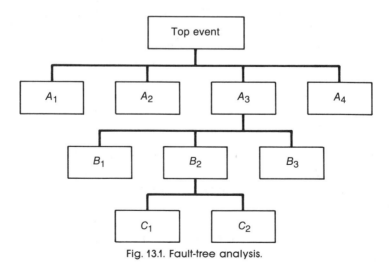

Fig. 13.1. Fault-tree analysis.

to equipment failures. They can also be human errors or changing environmental conditions.

The analysis is completed by assigning probabilities to each event and calculating the probability of the "top event" occurring using the Boolean algebra AND and OR conditions. If an event at a particular level cannot cause the effect at the next level without the assistance of one or more events at its own level, the AND condition is used: *A AND B* must both be positive in order to cause an effect at the next level. In this case the probabilities are multiplied. If any one of two or more possible events (at the same level) can cause the effect, then the OR condition is used: *A OR B* means that either one can be positive in order to cause an effect at the next level. In this case the probabilities are added. (In order for the OR condition to work properly, the probability of all possible events at that level must sum to 1.00.)

Fault-tree analysis can be used to compare probabilities of all possible "top events" (concentrate on those with the highest probabilities) or it can be used to compare sections of the same tree (called "cut sets"). A "cut set" is a probability calculation along one line, from top to bottom, of a fault tree. Again, more effort should be directed at controlling those events found in the "cut sets" with highest probabilities. If, in the analysis, a single event is found that can cause an effect all by itself (called a "single-point" event), it should be studied and controlled carefully (similar to a "bottleneck" operation on an assembly line).

If, in Fig. 13.1, either of the events C_1 and C_2 can cause effect B_2, and the probabilities are 60% and 40%, respectively, then the probability of B_2 occurring is 0.4(0.6) = 0.24 or 24%. Suppose that similar analyses had been made of B_1 and B_3 and they were found to have probabilities of 22% and 14%, respectively. Further suppose that all three of B_1, B_2, and B_3 must operate in order to cause A_3 (note that B_2 has now turned from effect to cause). The probability that A_3 would operate would be 0.14 + 0.24 + 0.22 = 0.60 or 60%. This entire analysis leading to the A_3 probability determination is a "cut set." Similar "cut-set" analyses could be made for A_1, A_2, and A_4 and their probabilities compared. Suppose that the probabilities of A_1, A_2, A_3, and A_4 are, respectively, 8%, 11%, 60%, and 18%. Obviously the A_3 "cut set" must be analyzed first for corrective and preventive action.

Procedure Analysis

FMEA, already explained, is a means of analyzing equipment failures. Procedures analysis is a method of using FMEA to analyze the effects of human errors. Basically, the procedures are the same except that the procedures analysis examines the actions that must be performed, the equipment that must be operated, and the environment of the workplace that must be maintained to ensure that the methods used are effective and efficient in the human–machine relationship.

The analysis examines the tasks, exposure to hazards, methods and procedural steps, equipment characteristics, and mental and physical demands. All matters that could degrade performance or cause injury, even though the prescribed method is used, are carefully examined for possible corrective and preventive actions. The outputs of procedures analysis are the same as FMEA — a list of procedural steps that might cause an accident, the effect or description of the accident, cause, and corrective or preventive action.

Contingency Analysis

If a hazard cannot be eliminated completely, the possibility always exists that an accident can occur. Contingency analysis considers the emergency measures that must be taken to avoid an injury or damage in case something does go wrong. Procedures are developed to cope with each contingency (each possible emergency). Emergency equipment is secured and properly located, escape routes are designated, rescue procedures delineated, and personnel are trained to cope with the possible emergency.

13.6 Design, Reliability and Safety

The modern concept of safety is that safety should be designed into a product to eliminate the possibility of a hazard instead of trying to control that hazard later with safeguards and/or rules of conduct. The problem with safety rules is that they attempt to control the human element in the human/machine interface by depending on the operator to obey a set of rules. In the past, this has ignored the fact that people are human with natural human weaknesses. They forget, daydream, do not notice a condition, react too late, to the wrong thing at the wrong time, do not react at all (freeze), etc. Sooner or later an operator is going to do one of these natural human things and, if the conditions are right, an accident will occur. Thus the importance of designing hazard-free equipment and products is apparent. Unfortunately, most products today were designed before this new concept of safety became generally known. In fact, even today, many (probably most) design engineers are not trained in safety attitudes and do not interface with the safety engineer in the design phase. There is still much to do to implement this eminently sensible idea into today's society.

One problem with this implementation is that it is likely to affect product reliability. Either some of the components of the product will have to be changed or new components added or both. Both of these types of changes will almost certainly decrease the reliability of the product. More complex components mean that the components are now less reliable, which in turn decreases the reliability of the entire product. The components now must be further engineered and changed in order to offset this decrease. If the components cannot be changed, it may be necessary to add redundancy in order to return the product to acceptable reliability levels. If new components have been added to a product as a result of safety, they will also decrease the reliability. (A new component itself has a reliability.) Once again, it may be necessary to further complicate the product by adding redundancy to increase the reliability until it has returned to its original value. All of these activities are going to increase the cost of the product, perhaps substantially. The challenge to the design engineer, then, is to somehow design safe and reliable products without substantially increasing the cost. Quite a challenge indeed!

Design engineers must keep a close lookout for hazards as they design a product. In order to do this effectively, they need to be trained in the principles and procedures of hazard recognition and control. They must understand the 10 different types of hazards, the adverse effects each can have on operators and workplaces, and how to recognize the inherent hazards of the design (the probability that a design will include one or more of the 10 types of hazards). Checklists are as important, and as useful, here as they are for the safety engineer. An excellent means of determining the inherent hazards in a design, besides the use of checklists, is the failure modes and effects analysis already explained.

At any rate, as soon as the hazards are identified, the engineers must be able to eliminate each type of hazard from their designs. The following list is just a partial summation of some of these hazard elimination procedures:

1. *Falls and impacts.* Slippery surfaces can be roughened with matte finishes. Rotating objects can be designed to eliminate any possibility of them breaking up (and thus becoming lethal missiles), or they can be designed into the center of the product to eliminate any possibility of them coming into contact with personnel. Just installing guards onto exposed moving parts is a less desirable alternative.

2. *Mechanical injuries: cutting, shearing, breaking, straining.* Remove all sharp edges, corners, burrs, and protuberances. Completely enclose rotating and ram-driven equipment. Use mechanical feeds where possible. Design guards and safety devices to be safe in all conditions with no access during operation, and as automatic as possible (no possibility of the operator to misadjusting it).

3. *Heat and cold.* Design heat and cold insulation into the product where needed. Do not depend on operating personnel to insulate themselves.

4. *Pressure.* Design safety margins into pressure vessels large enough to withstand any possible pressure increase, whether by a sudden

surge or by operating personnel error. Include fail-safe devices for fired pressure vessels to ensure that coolant water cannot be shut off. Include safety valves and blow-out plugs to relieve undue pressures, while making sure that they discharge to the rear or to areas where personnel are not likely to be working. Eliminate hose connections to minimize leaks and broken hoses. (Broken hoses can whip back and forth at a high speed and with great force causing much damage and injury.) Require enough testing — pressure and nondestructive — to eliminate failures due to structural defects.

5. *Electrical.* Eliminate all bare connectors or else contain them inside the product so that they cannot make contact with personnel. Specify insulation that will not break down under operating loads or under unexpected power surges and that will resist mildew, rot, and rodents for as long as the equipment can possibly operate. Make sure the equipment is protected and grounded such that operating personnel cannot receive a shock; double insulate wherever possible. Contain sparks, hot surfaces, and other ignition sources within the equipment to eliminate burns and fires. Protect equipment from inadvertent activation and deactivation with properly designed and positioned switches and buttons.

6. *Fires.* Contain all ignition sources within the equipment to eliminate inadvertent ignition of a combustible material. Separate combustible material from ignition sources and/or oxidizers. Include fire detection and alarm equipment wherever fire is possible and the possibility cannot be eliminated.

7. *Explosions.* Identify those materials that are explosive under the operating conditions of the equipment and specify special protective devices and procedures. Design so that explosive damage will be minimized, to both equipment and personnel, in case an explosion does occur. Provide for explosive material to be widely separated and in small lots.

8. *Toxic materials.* The design engineer must understand the effects of various types of toxic materials upon personnel and equipment and know which materials are normally toxic and which are toxic in combination with other materials. With the increasing use of newer and exotic materials, it is getting more difficult to keep up with their toxicity and damage possibilities. Design to contain toxic materials so that they can in no way come in contact with personnel. Study the effects on toxicity of combining materials and make sure that such combinations are not brought together under normal operating conditions. If possible, design to eliminate such combinations under any possible condition, no matter how bizarre.

9. *Radiation.* Basically, radiation damage (both the ionizing and the nonionizing types) can only be eliminated by shielding. It is far preferable to shield the source than to shield personnel, although both are usually needed for heavy radiation sources. Microwaves, a nonionizing radiation source, has just recently been recognized as dangerous, even in relatively low concentrations. This will be a continuing problem in the future, no doubt.

10. *Noise.* Check the product for sources of vibration and redesign to eliminate or reduce the vibration. Provide more secure foundations to prevent and absorb vibration. Provide more massive skeletal structure to withstand and dampen the vibration source. Noise insulation, that absorbs the vibration and turns it into heat, can be provided around the machine. Sometimes a secondary noise can be used to interfere with, and mask, the offending noise. Providing ear protection for personnel, although frequently necessary, is considered a less acceptable alternative to eliminating the noise source. People too often forget, or just do not wear the ear protection because it is uncomfortable.

14 | RELIABILITY MANAGEMENT

In this chapter, the management of a reliability function is discussed, its tasks listed and explained, and its interrelationships with product assurance and quality assurance are examined.

Objectives

1. Understand the elements of reliability planning and know how to accomplish a reliability program plan.
2. Know the tasks that need to be performed to implement a reliability plan during the product development stage.
3. Know the tasks that need to be performed to ensure reliability during the manufacturing phase.
4. Know the tasks that need to be performed to follow through on product reliability during the product's service life.
5. Understand the interrelationships of reliability, quality assurance, and product assurance, and know the tasks of quality control.
6. Know some of the organizational structures that are used to implement the reliability tasks.

14.1 Reliability Planning

Every reliability program should have a list of detailed procedures for accomplishing its tasks. This is the reliability program plan. It is of the utmost importance because it is the major vehicle which provides for efficient operation as well as being the medium (standard) by which the success of the operation is measured. The major elements of this plan are:

1. A detailed list of all tasks.
2. A description of each task with the proposed methods of accomplishing each task.
3. A detailed description of the method of evaluating the effectiveness of each task.
4. A master schedule showing the time each task is to begin and end.
5. Manpower and budgetary requirements.
6. Provisions for changes to the plan as requirements change.
7. A list of responsible departments and the tasks for which they are each responsible.
8. A list of known and probable reliability problems, an assessment of the probable problem impacts on the product, and the proposed solution procedures to these problems.
9. Provisions for reliability training.
10. A benefit–cost analysis to support the program and its budgets. The benefit–cost analysis should include, besides the cost comparisons, such intangible elements as customer satisfaction and company reputation, even if they cannot be quantified.

Every reliability plan should also include a program for improving the reliability of the product. This reliability improvement program should be designed to optimize reliability while at the same time reducing costs and increasing profits, increasing output without increasing unit costs, and increasing customer satisfaction. The first step in such a program would be to integrate the reliability and product assurance programs among all applicable company activities such as purchasing, engineering, research, manufacturing, quality control, inspection, packaging, shipping, installation, field-service, and performance feedbacks. Other important reliability improvement activities would include:

1. Make sure that products are designed to use those elements that optimize reliability and maintainability but, at the same time, keep life-cycle costs to a minimum.
2. Reduce the number of components.
3. Use better component arrangements.
4. Select better materials.
5. Use reliability checklists in all phases of the product life (design, development, manufacturing, and service life) to identify possible errors and to correct for them.
6. Debug and/or derate all components prior to use.
7. Minimize improper equipment use by providing for proper installation, good maintenance, and good operating manuals.
8. Provide warning labels, load and speed limits, and proper buttons and controls to minimize operator error.

9. Implement an information feedback, analysis, and control system that includes effective corrective action procedures.
10. Implement a failure modes and effects analysis program.

A reliability program depends on the product; on its complexity, how it is to be used, what its cost limitations are, and what the customer requires. A reliability program, then, would be deeply interested in, and should be deeply involved in, the design and development phase of a product. Correction of reliability design errors can be costly once the product reaches the manufacturing phase, and certainly after it begins to be used.

Actually, design limits reliability. A product can be no better than the bottom of the bathtub curve that describes its failure rate (see Chapter 1). Since manufacturing and service use tend to degrade a product, the actual reliability would tend to be even lower than the design limit. Thus, a good reliability program would be involved in the entire product life cycle, from design to development to manufacturing and, finally, to service use of the product.

The reliability function is a complex interrelationship of many simultaneous and sequential tasks; and the control of these tasks (to see that they are accomplished efficiently, in the proper order, and at the proper time) is not an easy job. This is the duty of reliability management. In order to better understand this complex function, reliability will be broken down into its component tasks with a description and definition of the tasks and a description of the procedures used to implement these tasks. The tasks are arranged in the order of a product's life cycle: development, manufacturing, and service use.

14.2 Product Design and Development Tasks

Reliability costs are minimized when reliability is designed into the product at its inception. The reliability program tasks that deal with and affect the design and development stage of product life are of utmost importance. The major reliability development tasks are:

1. *Requirements analysis.* Study the contractual requirements and specifications to make sure they are understandable and complete, and make recommendations for changes as needed. This task should be accomplished at the earliest possible date before the company gets completely locked in on unrealistic specifications that could, literally, lead to loss rather than to profit. This task would be best accomplished by making a complete list of the contract provisions and specifications, comparing them to applicable company specifications for similar products, noting where major variations could cause possible problems, noting what kind of problems are likely to result, and making recommendations to deal with them effectively.

2. *Company reliability specification preparation.* Prepare a reliability specification for the operation that would include the necessary procedures, measurements, and reports to properly control its implementation. The company specification should also include the customer desires as embodied in the contract provisions and its specifications, a statement of company policy, a statement of specification costs and possible tradeoffs to reduce costs and increase reliability, and a description of the product characteristics and functions. Design specifications define the reliability goals of the operation, guide the design engineers in their reliability efforts, and control suppliers reliability efforts. Specifications, of course, as was explained in Chapter 1, should be clear, unambiguous, understandable, and specific.

3. *Apportionment.* Prepare an apportionment schedule of reliability objectives by allocating the product reliability goals to its various subassemblies and then to the components and parts of each subassembly. Tradeoffs and adjustments can be made as development progresses in order to balance redundancy, safety, and cost considerations with the reliability goals. For example, suppose a product must have a reliability of 95% and that it consists of four subassemblies connected functionally in series. If three of the subassemblies are standard design, with their failure rates known, the reliability goal of the fourth, which must be completely designed, can be calculated. Suppose that the three standard subassemblies had reliabilities of 96%, 96.5%, and 99%, respectively. The reliability goal of the fourth, then, must be $0.95/(0.98 \times 0.995 \times 0.99) = 0.9841$ or 98.41%. If, as development progresses, any one or all of the subassemblies are changed in order to meet cost and/or safety considerations, the calculations of the reliability goals for the subassemblies would have to be redone. The important thing is that these early estimates will permit comparisons of various kinds of design alternatives as well as pinpoint areas of the design where redundancy is needed. Suppose the three standard assemblies were 96%, 97%, and 98%, respectively. Now the reliability of the fourth would have to be 104.1% in order to meet the contractual specification of the product (95%). Since nothing can be more than 100% reliable, one or more of the assemblies must have parallel components in order to have reasonable and feasible reliability goals.

4. *Reliability prediction.* In order to guide the design efforts, reliability estimates should be made early in the development stage. As was stated in an earlier chapter, redesign after the product is being manufactured is an expensive proposition. The feasibility of the reliability requirements, and any alternative design approaches would be evaluated and defined for their value to the overall design

objective. The risk of nonconformance would be determined as well as the cost of failures. In this way the design cost can be compared to the reliability early in the program. See Chapter 12 for a discussion of the various methods used to predict reliability at the design stage. (Also see Chapter 15.)

5. *Analysis of suppliers' facilities.* An analysis of suppliers' plants is done to see that their manufacturing facilities, and their quality and reliability plans, are adequate to support the product or contract requirements. This should include a study of the plant organization, responsibilities of key personnel, manpower, product-assurance manual, methods of selecting parts and material, equipment reliability, maintenance facilities, and proposed designs. Information on plant facilities can be obtained from a questionnaire designed for the purpose or from actual plant visits by reliability personnel. Both methods should be used if the project is very large or complex.

6. *Collection, analysis, correlation, and feedback of data are most important tasks of reliability.* These tasks require a high degree of statistical and mathematical skill on the part of the reliability professionals. Data from many different phases of the reliability program must be analyzed for completeness, adequacy, and accuracy, and then refined with mathematical, statistical, and empirical procedures to enable rapid interpretation. Incoming data are changed and summarized in this process to a form that is more meaningful and understandable to users (engineers, managers and designers).

7. *Design of tests and experiments.* Tests and experiments must be designed that gather the facts and also help direct efforts toward corrective action. These data can also be used to evaluate the efforts of engineers, designers, and suppliers. Many procedures are available to the reliability engineer to accomplish this task including reliability prediction models, stress analysis, derating techniques, end-of-life parametric analyses, design reviews with checklists, various types of statistical and mathematical models, reliability testing to determine failure rates and MTBF values, reliability overtesting to determine design limits, prototype testing, strife testing (stress and life testing), failure analyses, failure modes and effects analysis, and sequential testing. Although some of these procedures have been explained in earlier chapters, many of them are advanced models that are beyond the scope of this text.

8. *Evaluation of prints and specifications.* The proposed design, whether in-house or from suppliers of parts or components, should be evaluated for operational reliability, optimum combination of parts and components, simplicity in controls, ease of maintenance, compatibility with environmental factors (temperature, humidity,

vibration, etc.), and probability of human error. Reports of prototype tests should also be evaluated by examining such things as the test data analysis, corrective action reports, and recommended modification reports.

9. *Internal consultation.* The reliability department is in a unique position to provide consulting services to the design and development function. Reliability engineers are generalists with knowledge and experience in many functions, especially in the area of statistical and mathematical analyses. Thus they can be valuable in assisting the design function in solving reliability problems relating to manufacturing, purchasing, storage, packaging, and others, as well as those relating to design itself. This internal consulting service is not limited to design engineering, of course. All functions of the organization can benefit from this service as well as suppliers who have problems meeting their reliability requirements.

10. *Reports.* All functions must make reports in a well-run organization, and reliability is no exception. Reliability, however, owing to the investigative nature of most of its work, must submit many more reports than do most other organizational functions. Reports are the results of most of its activities and the object of many others. Administrative reports cover expenses, travel itinerary, conferences, program and activity status, periodic progress (weekly, monthly, and quarterly), plant surveillance, and reliability achievement. Technical reports cover failure rate analyses, problem areas and recommendations for corrective activity, plant reliability performance, plant schedule performance, reliability predictions and evaluations, engineering reliability requirements, test and evaluation studies results, and results of special projects. Reports are issued as required by the department and organizational policy, in accordance with customer requirements, or as a result of test and/or experiments.

14.3 Reliability during Manufacturing

If the project is large or the product is complex, a reliability engineer should be assigned full time, as a kind of internal consultant, to assist the design engineer during the development phase. Even for smaller projects and less complex parts, this type of assistance should be given, although it will likely be on a part-time or as-needed basis. At any rate, the engineer's familiarity with the product, and its possible problem areas, will be invaluable to the reliability function in its responsibility to manufacturing. The major tasks of reliability in ensuring a reliable manufacturing function are:

1. *Production readiness review.* Before production begins, the manufacturing facilities should be examined to confirm that the facilities

(plant, equipment, personnel, and methods) are all in proper condition to ensure a reliable product. Probably only critical processes would be analyzed along with acceptance test programs, analysis of parts and materials to see that they have been properly tested and certified, analysis of problems and their corrective actions to see that the problems have been properly resolved, verification of suppliers readiness, and examination of the control procedures to see that they are adequate to meet the reliability objective.

2. *Quality control review.* Quality control should be examined to see that procedures are adequate to meet the required quality objectives. This analysis should include a review of acceptance test procedures, product inspection and testing, defect classification and corrective action procedures, process controls, test equipment inspection and calibration, tool and gage proofing, and data gathering and analysis.

3. *Configuration control.* This is the proper control of drawings, product and process specifications, process procedures, test and inspection equipment, design, test procedures, inspection procedures, drawings and methods changes, and production orders and routings. It is important that these things be tightly controlled and that any changes be released formally (and controlled by the formality of the system). For instance, one person should be in charge of the drawings, see that all print copies are numbered, know where each copy is located within or outside the plant, see that all print changes are distributed to all locations, and see that all old copies are collected and destroyed. Similar contols should exist for all items listed above.

4. *Production reliability assurance.* Predicted reliability, estimated during development, tends to degrade during manufacturing, owing to slow deterioration in processes and procedures. It is the duty of reliability to provide assurance that this has not happened, or, if it has, to show that corrective action procedures have restored the part to its rated reliability level. Although this task is usually fairly simple, it is of the utmost importance. Usually a periodic check (mostly monthly) is made of a small sample of critical processes and procedures, which are inspected for changes that can decrease performance. If changes are found, corrective action procedures are instituted immediately. The procedures of this task are also used to qualify any changes in the configuration during the production cycle.

5. *Failure analysis and corrective action.* The reliability function should constantly check on the overall product and the individual component reliabilities. Reliability tends to degrade rapidly unless frequently checked and corrected. Failure analysis and corrective action procedures, as explained in Chapter 7 and elsewhere in this text, should be a constant, on-going, duty of the reliability function.

6. *Collection, analysis, correlation, and feedback of data are just as*

important, if not more so, during manufacturing as during devel-opment. The procedures for implementing this task are the same as for the similar task in the development phase. However, man-ufacturing data are drawn from different sources. Some of the most important are manufacturing plans, flow diagrams, shop orders, tool drawings, inspection plans, procedures analysis, process and product specifications, test and inspection data, and failure analyses. Although reliability may often perform their own tests to get their data, much of the data they use for their analyses come from pro-duction and quality control reports.

14.4 Reliability during Service Use

The responsibilities of reliability do not end at the manufacture and sale of the product. Estimated failure rates, predicted during the development stage, must be verified during actual use of the product in the field. Reliability field engineers are usually assigned to follow up on any large or complex operation by working closely with the installation and operation of the equipment in the field. Even if the product is not large or complex, follow-up information is obtained from the user so that failure rates can be verified. Some major reliability tasks during service use are:

1. *Service life evaluation.* This is a method of determining the remaining life of a unit being used in the field. Field failure analyses, trend studies, and accelerated aging tests are used to obtain the data. The information from this type of evaluation is used to determine spares requirements and to reassess user environments and user competencies.
2. *Surveillance testing.* This test is used to supplement service life evaluations. Periodic samples are drawn from stores and tested (both destructive and nondestructive tests are made) to detect incipient failure modes.
3. *Failure analysis and corrective action.* This is the same procedure as explained for development and for manufacturing, but it is applied here to service life evaluation. User failure rates are determined and subjected to analysis (see Chapter 7) after which proper corrective action procedures are implemented.
4. *Warranty evaluation.* Field data, from methods described above or from user information, are used to analyze warranty costs. The results of this action could have a profound effect on failure analysis and corrective action priorities.
5. *Collection, analysis, correlation, and feedback of data from service life evaluations is conducted much the same as it is done in prior stages and for the same reasons.*
6. *Maintainability studies.* Until a product is in actual use and begins

to require maintenance activities, its maintainability cannot really be determined. Maintainability (see Chapter 9) is the probability that a system can be returned to service within a predetermined period of time. Estimates are made, of course, at the design stage, but maintainability is one of the most difficult values to estimate closely. It must be verified in the field — in actual use. When a failure occurs, it is most important that the real cause is determined. Is it the fault of the product or associated equipment? Is it really a failure or just an apparent failure or even a mistaken failure? Is it caused by inexperienced personnel, the environment, or usage, or any combination of these three? The time and the methods of restoring the equipment to operation (the maintenance action) must be properly determined. Is maintenance accomplished by switching in, or installing, a standby unit? Is redundancy involved? All of these questions should be answered and data about them should be recorded. These tasks take an experienced and well-trained reliability engineer to accomplish properly. Since he or she will be in close contact with customer personnel, interpersonal relations are critical. The field service engineers, in this type of assignment, represent the company and, as such, can do much to advance customer relations. Maintenance information is used to determine the maintenance action rate and thus the maintainability figure (see Chapter 9). Maintenance information is also useful in determining spares requirements.

14.5 Quality, Reliability, and Product Assurance

Both quality assurance and reliability personnel are responsible for studying the product for quality. Quality assurance is concerned with the ratio of defective product (how many parts out of a lot that do not meet the product specifications), while reliability concentrates of failure rate analysis (quality over the long run). Both groups use sampling techniques, testing procedures, and inferential statistical analysis to achieve their goals. There is frequently little difference between the techniques of quality assurance and those of reliability assurance. In fact, reliability will frequently use data gathered by quality control in their analyses. The same type of overlapping expertise is also present in a comparison of their personnel. Reliability and quality engineers receive much the same training (there is some variation but it is minimal) and experience, especially in relation to statistical and mathematical calculations. Therefore, both reliability and quality tasks are often found being accomplished by the same personnel, usually in small and medium-sized organizations.

Owing to the overlapping and interrelated nature of their duties, quality and reliability personnel should work together as a team. This marriage of the two functions is accomplished by product assurance.

The product assurance department should include all of those functions of reliability and quality which will assure a high degree of reliability in the product. Product assurance should report to top management and have responsibility for reliability surveillance in all those functions that affect reliability, such as design, purchasing, production, field installation, field service, and maintenance.

In the normal development of these two functions (quality and reliability), they have been separated in the past into two different departments. The modern concept, however, is to reunite them under the common goal, and function, of product assurance. This requires that the new product assurance engineers be equally knowledgeable in the tasks of reliability assurance (already explained) and in the tasks of quality assurance. Some of the more important tasks of quality assurance are analysis of product quality requirements and specifications; supplier and subcontractors plant surveys; design of company quality specifications; quality review of prints, specifications, manufacturing methods, etc.; plant surveillance and evaluation; consulting services; collection, analysis, correlation, and feedback of data; and reports and recommendations on quality problems.

14.6 Reliability Organizational Structures

There are three main methods of organizing the reliability function. The first method is the total separation of the two functions, where each is a separate department in the company (see Fig. 14.1). In this type of reliability organization, quality is not even included except as a liaison function. Quality and reliability would still interact, but in the more formal manner that two separate departments always interact in a company. There is no product assurance function.

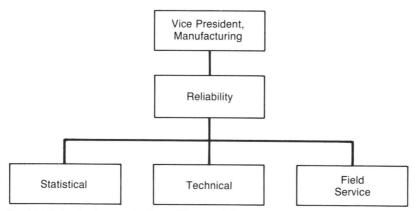

Fig. 14.1. A typical organization with completely separate reliability and quality functions.

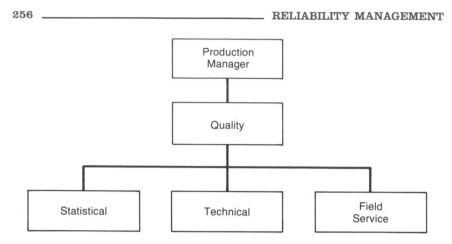

Fig. 14.1—*Continued*

The next organizational structure is designed for a product assurance function where reliability and quality, still separate, report to the same person, the product assurance director (see Fig. 14.2). Although the two functions are still separate, they do belong to the same department, which allows a considerable latitude for informal, as well as formal, interaction.

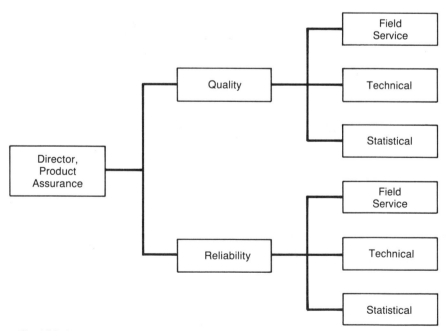

Fig. 14.2. A typical product assurance organization with separate reliability and quality functions.

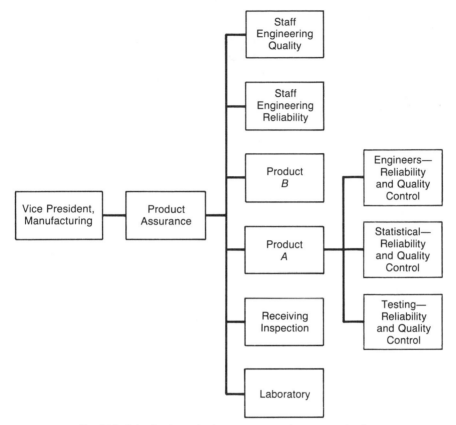

Fig. 14.3. A typical product assurance systems organization.

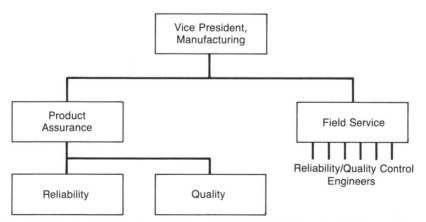

Fig. 14.4. A typical product assurance organization with integrated field operations.

The final structure is a systems approach where the quality and reliability functions are integrated on the task level. (See Fig. 14.3). It is in this more modern structure where product assurance engineers, trained in handling both quality and reliability problems, replace the traditional quality and reliability engineers.

A fourth organization (see Fig. 14.4) is a cross between separate functions and the systems approach. In this structure, the field service engineers represent both reliability and quality to the customer. It makes good sense to combine this function in this way since the field service engineer represents the company to the customer. Too many representatives can be confusing to the customer.

CHAPTER 15 | PRODUCT LIABILITY

This chapter discusses the concept of product liability; the responsibility to make the product safe for the user. The proliferation of product liability lawsuits, as well as the increase in monetary awards by juries, emphasizes the great need for safety in design.

Objectives

1. Have a knowledge of the major events that led to the modern product liability concept and of the fundamental principles involved.
2. Understand the fundamentals of hazard recognition and analysis.
3. Know the models used to analyze a product for hazards.
4. Know some of the more important methods of designing safety into a product.

15.1 Product Liability Principles

Product liability refers to the safe use of a product where the product was used in a manner consistent with its design criteria (used in the way it was intended to be used). If injury or damage results from this use, the product (and the manufacturer) is liable for the damage (financially liable to pay for the cost of injury and/or damage).

Basically, injury or damage results from one or more of the following four causes: a material failure, a human error, an adverse product characteristic, or interacting environmental conditions. Usually more than one cause is operating, such as a product made of a material that cannot withstand subzero temperatures that is inadvertently left outside in the winter causing the material to crystallize leading to breakage at a critical time.

In order to have a safe product, all of these factors must be considered, and dealt with, during the design stage. Unfortunately, two fundamental human attitudes interfere with the control of the human factor. There seems to be an ingrained resistance to safety devices, especially if they are at all restrictive to the operator. And there appears to be a strong feeling, especially among managers, that safety programs can never be justified costwise — they always add more costs with little or no return.

With the steady increase in injuries and damages through the years of the industrial revolution, it has become apparent that a different attitude must be cultivated. Modern safety analyses, for instance, conducted with different attitudes and less bias, have found that most accidents could have been prevented by good design and planning (accidents are not caused by human error as has always, heretofore, been supposed). The need, in modern society, for product liability and accident prevention programs has been well established (see Chapter 13).

Privity Rule

It has always been accepted, under the common law, that manufacturers are responsible for the safety of others; for the safety of those in the plant (workers and visitors) and for the safety of the consumers who use their products (within certain limits, such as, the products must be used in the way they were intended to be used). However, with the increasing complexity of modern business practices, the ultimate user has been increasingly removed (in the sales chain) from the manufacturer. The manufacturer sells to the wholesaler who sells to the dealer who sells to the user, making it difficult under the law to assign responsibility for product failure. In 1842, in the case of Winterbottom v. Wright, the courts established the privity rule, which stated, in effect, that legal responsibility for product failure was assigned only to the next person in the sales chain. No liability was accepted unless there was a direct relationship between seller and buyer. The user could only sue the dealer from whom the product was purchased; the dealer could only sue the wholesaler; and only the wholesaler could sue the manufacturer. (This was because each was a separate business entity under the law. If the wholesaler and the dealer were direct employees of the manufacturer, the rule of privity would not be applicable, and the user would be able to sue the manufacturer directly.)

This condition led to too many injustices under the law. The user frequently could not collect from the dealer because the dealer was not responsible for the defective product and could not be expected to know of the defect, or else the dealer did not have the money to pay. Even in these situations, the user could not recover from the manufacturer because of the privity rule. (The dealer or wholesalers also could not recover for damages from the manufacturer because neither one was the injured party.) Finally, in 1916 in New York, this condition was corrected. Judge Cardoza, in the case of McPherson v. Buick Motor Company, ruled that the manufacturer has a basic responsibility to build safe products and to inspect these products to ensure their compliance to safety principles. This responsibility was also extended to cover parts purchased from subcontractors. The manufacturer was ruled to be financially liable and the privity rule (which really did not apply anyway to the changed conditions of modern business practices) was disallowed. The user could now sue the manufacturer for damages caused by faulty products.

Warranty

Another principle of product liability is the concept of warranty. There are two kinds of warranty: expressed or implied. An expressed warranty is a specific promise of the manufacturer, where legal responsibility is accepted (either written or oral) for product failure even if no injury or accident results. These types of warranties are always limited as to time, and are dependent on reliability predictions. If the reliability prediction is wrong, the manufacturer can lose a lot of money. An excellent example of this was the 50,000 mile automobile warranty. Although reliability calculations predicted a low enough failure rate to justify the warranties, actual experience proved them wrong. The 50,000 mile warranty proved too costly to continue.

Implied warranties, on the other hand, are imposed on the manufacturer by the courts. In effect, the courts say that the very act of offering a product for sale implies that it is reasonably safe to the user (Henningsen v. Bloomfield, 1959). The centuries old concept of *caveat emptor* (let the buyer beware) was disallowed, at least as far as user safety was concerned. Unlike the expressed warranty, this warranty has no time limit and is not related to reliability. In fact, it has been found that a product can have a high reliability and still have a characteristic that is dangerous to the user.

Strict Liability

In 1963, the California Supreme Court made another landmark decision that has since been adopted in the courts of many other states. In the concept of strict liability, or liability without fault, neither warranty nor negligence on the part of the manufacturer has to be present, or even be proved by the user. All the user has to do is prove that an injury occurred and that he or she exercised the same care as would a reasonably prudent man working in the same circumstances (the doctrine of "reasonable prudence," the definition of which has caused much controversy). The manufacturer is then considered to be financially liable unless he or she can prove otherwise. The burden of the proof has been shifted to the manufacturer.

Negligence

Most liability claims are won on the basis of negligence. By negligence is meant that the product was designed, produced, installed, or maintained in such a way that injury or damage occurs. Sometimes a product must be produced in such a fashion that a dangerous or hazardous condition must continue to exist. (The nature of the product and its intended use make it impossible to eliminate the hazard.) In this case, even if everything else is correct, negligence is still present unless a suitable warning is properly presented to potential users.

Several conditions have been identified that can be accepted as proof of negligence. Failure to anticipate all possible uses, or even misuses, and allow for them is one condition that can lead to a claim of negligence. There is much controversy, however, in what actually constitutes a "failure to anticipate." This is too frequently a matter of opinion, leading to many controversial court decisions. Another condition that can lead to a negligence claim is if the manufacturer failed to use "due care" in order to foresee a potentially hazardous condition. This type of claim, once again, is open to interpretation leading to conflicting decisions in the courts. One of the best ways to prove negligence is to show that good manufacturing or design principles were not followed. If any exceptions can be found, it is up to the manufacturer to show good reason why the good practice was not followed. (There are often excellent and justifiable reasons for deviating from accepted practices.) This requires that complete records be kept. One of the problems with proving a poor practice claim has been the difficulty of getting these records. In the past, the plaintiff has had to be quite specific as to which record was desired before the courts would grant access to it. However, the modern concept of "discovery" has corrected this inequity. The plaintiff can now request all records in a particular category, and the defendant is obliged to turn over all documents that a reasonable person would ask for if aware of them. A careful attorney can usually word the request so that it includes almost all documents available.

Defenses Against Liability Suits

A good safety effort and the records to back it up can be instrumental in defending a manufacturer (or anyone) from liability suits. This should include:

1. Good records, with all the facts included and logically organized.
2. An effective and consistent safety program.
3. Competent and well-trained safety employees.
4. A competent, well-organized, and up-to-date hazards recognition and control program.
5. Prompt correction of each hazard.
6. Provisions for safety devices on all hazards that cannot be eliminated.
7. Warnings clearly displayed on signs, decals, operating instructions, etc. (when a hazard cannot be eliminated).

15.2 Analyzing Hazards

In order to control or eliminate a hazard, it must first be recognized and then analyzed. Recognition of a hazard comes from experience related to the product, or to other products, and by various types of analyses. The best way to ensure that a familiar hazard is recognized is by the use of checklists. Checklists are just lists of questions that draw attention to possible hazardous conditions that have been analyzed in the past by safety professionals. Hazard analyses are formal, step-by-step, procedures that examine the elements of products, operations, or systems in order to identify conditions that might be able to cause accidents, injuries, or damage, and then to identify and institute procedures for eliminating or neutralizing the effects of these hazards. There are many factors involved in these analyses that must be understood by the analyst.

Types and Categories of Hazards

There are three types of hazards: initiating, contributory, and primary (sometimes called critical or catastrophic). An initiatory hazard is a condition that starts a chain of circumstances leading to eventual failure. Contributory hazards then combine with the initiating hazard and eventually lead to a particular type of failure called the primary, or catastrophic, hazard. This is a process where one hazard combines with a second one to cause a resultant condition that can then combine with a third hazard causing another resultant condition, and so on, until a failure finally occurs. The failure is then a primary hazard if it causes injury or death, damage to equipment or buildings, degradation of the function (lessened ability of the equipment to perform its intended function), or loss of material (leaks, spills, etc.).

Accidents and/or hazards should not only be analyzed as to type (initiating, contributory, and primary), but also by category. First of all, hazards can be inherent in the physical properties of the product. A pressure vessel made of plain carbon steel which can rust (and eventually rupture) is an excellent example of this category. Material or human failure is another category of hazard. A pressure vessel designed with material too thin to sustain operating pressures would be an example of this category. If the pressure vessel is subjected to moisture, another category of hazard is identified — environmental stress. Usually these categories and types are combined in any one product making analysis difficult and accident prevention costly.

Human Factor

The human factor leads to many accidents. In the past, most accidents were thought to be caused by human error. Now it is recognized that design deficiencies cause most accidents, even when a human error is present. (Too often, designers ignore normal human weaknesses in their designs making it possible for errors to occur.) Of course, humans make the design errors so that it can be said that most accidents can be traced to human error. Designers must be trained to recognize where human error can cause problems and design to eliminate the possibility of human error. Some of the important categories of human error are:

1. Failure to perform an important function (an error of omission).
2. Performing an unnecessary function (an error of commission).
3. Performing a function out of sequence; doing the right thing at the wrong time (an error of timing).
4. Failure to recognize a hazard (an error of training or experience).
5. Inadequate response to a signal (an error of training, experience, or fatigue).
6. Wrong decisions (an error of judgment).

Design errors can also be categorized for analysis (the act of categorizing an error frequently suggests the method of eliminating it):

1. The inclusion of materials or conditions that are obviously incompatible with each other or with the product's intended environment.
2. Making a product technically practical but incompatible with intended operating conditions.
3. Making a design, or specifying operating procedures, that violate normal human tendencies or expectations (number from bottom to top or counterclockwise).
4. Placing undue stress on operators (noise, heat, burdensome protective equipment, undue strength requirements, unusual positions, etc.)

Other categories of human errors are those that occur during the production of the product, during maintenance and repair, and during the operation or use of the product. Errors at any one of these places can totally negate good design. Therefore, designers need to be knowledgeable in these areas and specify proper production, maintenance, and operating procedures. Finally, the human–machine relationship must be considered. There are certain things a human can do better than a machine (where judgment and motivation are concerned) and certain things a machine can do better than a human (where strength and error-free repetition are concerned).

Hazard Analyses

A hazard analysis is defined as the investigation and evaluation of the following:

1. Interrelationships of the different types of hazards.
2. Factors involved in the safety of a product (operating conditions, equipment, personnel, and environmental conditions).
3. Means of eliminating or neutralizing the effects of a specific hazard.

The different types of hazard analyses can be categorized as follows:

1. Determine the theoretical causes and relationships and compare to similar products and past experience. Checklists are used extensively with this category.
2. Determine hazards and their causes from predesign and postdesign analyses. Obviously predesign is before-the-fact and postdesign after-the-fact.
3. Various quantitative analyses, which include the relativistic systems where similar conditions are compared mathematically; safety factors and safety margin systems where strengths were increased to withstand any possible stress; numerical ratings where expert estimates are made on a numerical scale and hazard ratings are added to give a numerical scale; and probabalistic analyses where probabilities are computed as in reliability models (Chapter 7).
4. Compare to models of the real world (the models represent some aspect of the real world). Iconic, or visual, models are usually smaller examples of the real thing, as in model airplanes. Analog models use one set of properties to compare and analyze a different set (using water flow properties to explain electricity). Finally, symbolic models use mathematical equations or computer programs or both to determine scale levels and probability values.
5. Safety analysis tables present safety factors in an organized manner that allows for ease of analysis. Possible hazards are identified and listed along with their causes, effects, and corrective action procedures.

15.3 Hazard Recognition Models

The various stages of the life cycle of a product contribute in different ways to the hazard characteristics of that product. Therefore, different types of models must be used for each life cycle stage when attempting to recognize, and provide controls for, product hazards. The models are designed to meet the peculiar objectives of each of the life cycle stages. At times, more than one model can be used to fulfill a specific objective. Also, many of the models have quite similar characteristics and procedures, the difference being the purpose and direction of the analysis (see W. Hammer, *Handbook of System and Product Safety*, Prentice Hall, Englewood Cliffs, NJ, 1972, for a more complete discussion of these models).

Preliminary Hazard Analysis

The purpose of this analysis is to identify possible hazards in a product or system where there is little similarity to other products or systems and where little or no experience exists. Because of this, the model should not be expected to identify all the hazards and it should not be supposed that all of the possible hazards identified will actually be present in the product. This model is used basically as a guide to the design and development function. Other, more detailed models must be used later as a supplement to complete the total hazard analysis. The procedure is to examine the product, similar products, documentation (contracts and specifications), and the performance requirements to determine the possibility of hazards. These hazards are then categorized into initiating, contributory, and primary, and are listed on a form. Then the cause of each hazard is listed along with its effect (or effects) on the product. Finally, the corrective or preventive measures are also listed giving a single form with all of the information needed for proper hazard recognition and control. Checklists are extremely useful in this model (and in most models) to assist in identifying the hazards and to guide the analysis.

Criteria Review

The purpose of this model is to review specifications, standards, codes, and safety regulations to avoid situations that, in the past, and with similar products and systems, showed indications of possible safety problems. Such specifications and codes can provide guidelines for design and development, information on possible hazards, a means for determining safety work loads, a standard for comparing actual performance, and assist in analyzing documents for costly or unrealistic specifications. The procedure is to determine the possible hazards in the product by an examination of the applicable documents, prepare sets of checklists to guide the corrective action analysis (checklists may already be available from previous studies), and the establishment of design standards from this analysis.

Mission Profile Analysis

This model is limited to analysis of a system. The system and subsystem functions are identified, along with the operating conditions, and then analyzed for their contribution to the mission, or purpose, of the system. This is an all encompassing model that includes analyses of design, manufacturing, transportation, storage, handling, maintenance and repair, operational use, and all support functions. All conditions of design, procedures, operations, and environment are examined to ensure that all support functions will be properly integrated to support the mission, the mission will be successful, and injury or damage does not occur. The procedures are:

1. Examine all documents to determine the system and subsystems functions.
2. Analyze the system and subsystem functions and their interrelationships to determine the mission tasks.
3. Make a profile of the mission by listing the tasks, responsibilities, and task functions, and how each function relates to the overall mission and all inherent problems associated with each task. Profiles also frequently include graphs of essential elements for quick and easy analysis.
4. Assign probabilities to each problem. If information already exists from previous systems, the probabilities can be computed. Otherwise they must be estimated by design and reliability personnel.
5. List preventive and corrective procedures for each problem.
6. High probability tasks can be further subdivided and analyzed as above, with this division continuing down step-by-step until it accomplishes the level of control desired.

Mapping

Mapping is a method of determining optimum relationships due to relative locations of the essential elements of the system in relationship to potential hazards. Various maps and layouts are studied to determine which one best limits the safety hazards. Examples of mapping include:

1. Determine the best method of separating fuels and oxidizers to ensure safety and optimum production at the same time.
2. Determine optimum arrangement of fuel lines and ignition sources.
3. Determine danger areas and optimum arrangement for personnel, structures, and equipment.
4. Map the hazard contours. The contours describe the areas where specific types of protective equipment will be needed (as in ear protection for a jet engine).
5. Failure plotting for failure analysis. The map analyzes the probabilities of different types of failures at different locations as well as the effects of the failure along a projected path of movement (train, aircraft, missile, etc.).
6. Analyze bus routes to determine the safest route to take. Hazards are noted and preventive procedures are described for each hazard area.

7. Analyze optimum emergency routes, safety zones, fire zones, fire defense routes, etc.
8. Meteorological analyses.

Environmental Analysis

Environmental analysis can be a part of other models or a model of its own. It consists of analyzing environmental factors to determine the type and extent of hazards and then providing for their elimination or control. A mission analysis may have to be done first in order to determine the probabilities of the system interaction with particular environmental hazards. A mapping analysis is also frequently done (often as part of a mission analysis) in order to analyze location interrelationships. Some examples of environmental stresses include:

1. Solar radiation.
2. Pressure (at different altitudes and in different pressure vessels).
3. Moisture (flooding and corrosion).
4. Wind (updrafts, downdrafts), prevailing winds, day/night differentials.
5. Wildlife (birds in vicinity of aircraft).
6. Land conditions (dry, wet, variable).
7. Smog.
8. Heat from large concrete masses.
9. Vibration and shock.
10. Artifical environments (oxygen and hydrogen, nitrogen, etc.).
11. Long-term effects (corrosion) versus short-term effects (flooding).

Subsystem Analysis

Subsystem analysis is actually a misnomer as it actually analyzes a system at all three levels of action, not just at the subsystem level. Hazards are identified at the subsystem level, at the interaction of one subsystem with others, and at the total system level. Different models are used at each level in order to properly identify each type of hazard. Subsystem analysis, then, is a mixture of several models joined together by a systematic procedure for analyzing complex systems. The procedures are:

1. Do a mission analysis to identify the system goals and objectives.
2. List the system hazards identified by the mission analysis.
3. Do a flow analysis to identify the subsystems. At times, subsystems are further broken down into their components and the components broken down into their parts. The amount and level of detail depends on the amount and type of information needed.
4. Do a procedures analysis and list the hazards thus identified.
5. Do a failure modes and effects analysis (FMEA) to identify possible failure points.

6. Do an interface analysis to identify hazards at subsystem inter-relationships.

7. Prepare preventive and corrective measures for each of the hazards. The output of this model is quite similar to the output of the procedures analysis.

Energy Analysis

In this model the assumption is made that all accidents are caused by uncontrolled, undesirable, transfer of energy. Therefore, an analysis of the sources and flows (transfers) of energy is made to determine hazards. There are many forms of energy analysis depending on the forms of energy, methods of transferring the energy and means of controlling it. The TNT equivalency method, for instance, compares massive transfers of energy to equivalent amounts of TNT (as if an amount of TNT were used that produces an equivalent amount of destruction). This is the method that is used to describe nuclear destruction, for instance (equivalent to 100 tons of TNT, etc.). Energy analysis can also be used to analyze heat transfer characteristics. Problems such as proper insulation needed, heat absorption, combustion possibilities, electrical losses, and equipment degeneration and failure can be solved with this model. After the energy transfer and flow characteristics are determined, and possible hazards are identified, methods are devised to control the transfers, reduce the flows, isolate the energy sources, or substitute or change the energy source to a less violent one.

Flow Analysis

Although flow analysis can be used as an independent model, it is mostly used to assist in locating hazards in the interface model just described. Water, heat, electrical, and energy flows are analyzed for hazards. There are two types of flows that can cause problems; confined and unconfined. Confined flows, such as water flowing through a pipe, cause the most difficult problems. The worst hazards, causing the most damage, are found in these confined flows. Unconfined flows, on the other hand, usually pose few problems. Heat transferred through the air is an excellent example of this type of flow. Unless the heat source is very large, great damage usually does not occur. Once the flows are mapped, the next step is to identify the characteristics of the flows (loss of pressure, corrosion, toxicity, flammability, etc.), the interface characteristics (with other flows or components), and the hazards that might be caused by these characteristics. In addition, connections for confined systems are examined for possible hazards. The hazards are then listed along with the cause, effect, and preventive/corrective measures.

Prototypes and Mockups

Prototypes are actual models of the product or system. Mockups are three-dimensional replica, not a working model as is the prototype, used as visual aids for analysis. The prototypes are used for extensive testing, prior to actual production. A mockup, on the other hand, is used near the beginning of the design phase to guide the development.

Human Engineering

Human engineering is concerned with the human/machine interface, with machine designs that minimize strain on the operator. The goal is to integrate the best capabilities of humans and machines for optimum operating conditions that minimize the chance of an accident. Examples of designs that increase operator strain are having too many instruments to watch at one time, widely separated controls, levers and wheels that require excessive operator strength, instruments that require constant monitoring (cannot look away), instruments numbered counterclockwise or with low numbers up and high numbers down, excessive heat and humidity, welding operations, and environmental conditions that affect the body psychologically as well as physiologically. The procedure is to review what people do best and what machines do best (a checklist is extremely helpful here) and the machine/human interface, and then match these abilities to the system requirements. Next note the problem areas and deficiencies in the personnel subsystems and take measures to eliminate or control the hazards thus identified.

Link Analysis

A link is any connection (communication such as visual, tactile, or auditory) between elements (components such as person, control, display, station, etc.) of a system. Link analysis is concerned with analyzing these connections for adequacy, load rate, position, arrangement, and frequency of interchange. The purpose is to arrange the elements of a system so that the most critical links are reduced to minimum length. The procedure is to make a drawing of the system showing the elements in position and to scale (unless distance is unimportant) with the communication links identified. Next, a study is made to determine the frequency of interchanges along each link and each link is rated, in relation to its frequency or use, on a scale of 1 to 10. Then, the importance of each link is estimated in relation to its difficulty, need, speed, and control requirements, and total time used, and rated on a scale of 1 to 10. The two ratings are then added together to determine the criticality of each interchange (each link). The link with the highest number is the most critical. Finally, the system elements are rearranged so that critical link distances are reduced to a minimum. Reducing a critical link distance usually requires that a link of lesser criticality is increased. Care must be taken that this does not also increase the lesser link criticality.

Simulation

Simulation is a technique to determine results of specific inputs into a system by the use of a model of that system. The model can be a mathematical model (such as fault-tree analysis), a computer model (which usually includes a mathematical model in its structure), or a three-dimensional representation of the actual system. Mathematical and computer models are usually limited to the early stages of a project, while three-dimensional models are more likely to be used during the later stages when some hardware is available. Examples of three-dimensional simulations are flight simulators to train aircraft pilots in emergency techniques, wind tunnels for testing aircraft prototypes, simulated environment testing, dummies in autos and aircraft, scale models of houses for fire testing, old houses ready for demolition for fire testing, and dummies for fire testing and parachute testing. Simulation as a device for safety testing has some limitations. There are just too many ways that an accident can happen and simulation depends on anticipated events.

Interface Analysis

The purpose of this model is to locate and deal with the hazards that exist as a result of interrelationships among the components of a system. It is important that these hazards are not allowed to cause accidents or impair the functions of other components (or of the entire system). In these interfaces there are two fundamental types of problems. Physical incompatibilites are caused mostly by dimensional inadequacies that lead to problems between mating parts, inadequate clearances for moving parts, and lack of capacity (holding tanks, electrical circuitry, etc.). Functional inadequacies are caused by mismatching of energy relationships between outputs and inputs. In complex systems the outputs of one component become the inputs to another. When they do not match properly, problems result.

Most complex systems have an enormous amount of interfaces. Identifying all of these interrelationships is the first step in applying this model. One of the best methods of identifying these interfaces is to sketch the system in a simple block diagram. The next step is to identify and categorize the hazards at each interface. Some of these hazards are obvious, some can be readily identified through the past experience of the engineers, and others found by combining this model with others already explained such as FMEA, time sequencing flow analysis, and network analyses. The final steps, as in most of these models, consist in determining the type and extent of possible damage that can result from each hazard and means of eliminating, reducing or isolating these effects.

Other Analyses

Earlier chapters detailed other methods used in hazard analysis: criticality analysis, Chapter 13; procedures analysis, Chapter 12; contingency analysis, Chapter 13; fault-tree analysis, Chapter 13; and FMECA, Chapter 12.

15.4 Safety Design Methods

The final result of a hazard analysis is the determination of methods of eliminating or controlling the hazardous conditions. This text cannot begin to cover all the methods available to a design engineer to control hazards. For one thing, each type of hazard has its own control methods and, for another, much of the elimination and control of hazards depends on the ingenuity of the designer. However, there are some tried and true principles that have been developed through the years.

The lists of design safety methods are growing daily. It is incumbent upon the design engineer to become completely familiar with these methods and to use them as needed to increase product liability. Some of the most important methods are listed below:

1. Redesign to eliminate the hazard; by far the most desired action.
2. Provide clear, concise, and understandable instructions.
3. Provide warnings, cautions, and explanations for those hazards that cannot be eliminated.
4. Minimize stress judgments by preprogramming contingency measures.
5. Design to eliminate improper installation by such things as irregular hole patterns and different size connectors.
6. Make recognition and warning patterns match expectations (clockwise instead of counterclockwise, for instance).
7. Make sure response time is adequate and matches human capabilities.
8. Avoid interference, awkward controls, or similarity between controls.
9. Provide backout and recovery procedures to abort the operation, or restore normal sequencing as needed, when errors or unusual conditions occur.
10. Reduce the failure rate with safety margins derating and redundancy.
11. Provide fail-safe designs to shut down (fail-passive) or continue operation (fail-active) in emergencies.
12. Provide for isolating the effects of a hazard when an accident

occurs by distance (explosion) or by absorbing, deflecting, or containing the energy release.

13. Provide personal protective equipment for dangerous situations.
14. To prevent inadvertent activation, provide torque-type switches, recessed switches, guards over switches, interlocks, lockouts, controls in the sequence in which they are used, clock-type controls to eliminate need for fine adjustments, etc.
15. Protect against hot surfaces, live wires, sharp objects, and hard surfaces.
16. Provide warnings (loud sound or light) for emergencies (to attract operator's attention).
17. Avoid operator stress by eliminating tiring requirements, excess loads, long concentration times, vibration, and awkward positions.
18. Provide adequate reference points to minimize fixation or disorientation.
19. Provide adequate environmental controls. Prevent inclusion of heat, moisture, or contaminants. Provide adequate heating or insulation. Provide proper life support equipment.

REVIEW
OF
ALGEBRA

A thorough knowledge of the following rules, definitions, and algebraic manipulations is necessary for a proper understanding of this text.

Rules

1. To add two numbers with like signs (both positive or both negative) add their absolute values and prefix their common sign:

$$-4 + (-6) = -10 \quad \text{or} \quad +4 + 6 = +10$$

2. Too add two numbers with unlike signs, subtract the smaller absolute value from the larger one and prefix the sign of the numerically larger number. If the numbers are numerically equal, their sum is zero:

$$
\begin{array}{r}
-10 \\
+\ 5 \\
\hline
-\ 5
\end{array}
\qquad
\begin{array}{r}
+11 \\
-10 \\
\hline
+\ 1
\end{array}
$$

3. To subtract one number from another change the sign of the number subtracted and then add:

$$6 - (-5) = 6 + 5 = 11$$

$$6 - (+5) = 6 - 5 = 1$$

4. The product and the quotient of two numbers with like signs are positive; of two numbers with unlike signs, negative:

$$4 \times (-6) = -24 \qquad -24/+6 = -4$$
$$-4 \times (-6) = +24 \qquad -24/-6 = +4$$

5. The value of a number is unchanged when 0 is added to it or subtracted from it:

$$6 + 0 = 6$$
$$6 - 0 = 6$$

6. The product of any number and zero is always zero:

$$6 \times 0 = 0$$

7. The quotient obtained by dividing zero by any number is zero:

$$0/6 = 0$$

8. It is impossible to divide any number by zero:

$$6/0 = \text{undefined}$$

9. The sum of two or more algebraic terms is not changed by changing their order:

$$X + Y = Y + X$$

10. The sum of any group of algebraic terms is the same, however they are grouped:

$$x + y + z = (x + y) + z = x + (y + z)$$

11. The sum of the products of one term by each of several other terms is the product of this term by the sum of the others:

$$AB + AC + AD = A(B + C + D)$$

12. When parentheses preceded by a minus sign are removed, the signs of all terms which had been inside are changed:

$$4 - (x + y) = 4 - x - y$$
$$4 - (x - y) = 4 - x + y$$

13. The product of two or more factors is not changed by changing their order.

$$2 \times 3 = 3 \times 2 \qquad AB = BA$$

14. The product of three or more factors is not changed by grouping them in different ways:

$$(AB)C = A(BC) = B(AC)$$

15. Laws of exponents:

$$A^m A^n = A^{m+n} \qquad \frac{A^m}{A^n} = A^{m-n}$$

16. Identities:

$$A(B + C) = AB + AC$$
$$(A + B)^2 = A^2 + 2AB + B^2$$
$$(A - B)^2 = A^2 - 2AB + B^2$$
$$(A - B)(A + B) = A^2 - B^2$$

17. The sign before a fraction is changed when the sign of either the numerator or the denominator is changed:

$$\frac{-6}{7} = -\frac{6}{7} = \frac{6}{-7} = -\frac{-6}{-7}$$

18. The sign before a fraction is changed when the sign of a factor of either the numerator or the denominator is changed:

$$+\frac{-6(7)}{8} = -\frac{-6(-7)}{8}$$

19. The product of two or more simple fractions is the product of the numerators divided by the product of the denominators.

$$\frac{A}{1} \times \frac{B}{C} \times \frac{D}{E} \times \frac{F}{G} = \frac{ABDF}{CEG}$$

20. To divide one fraction by another multiply the first one by the reciprocal of the second.

$$\frac{A}{B} \div \frac{C}{D} = \frac{A}{B} \times \frac{D}{C} = \frac{AD}{BC}$$

21. If $\log_a b = d$, then $b = a^d$:

$$\log_{10} 100 = 2 \quad or \quad \log_{10} 10^2 = 2 \quad or \quad 100 = 10^2$$

Definitions

1. $$A^0 = 1; A \neq 0$$

2. $$A^{-k} = 1/A^k$$

3. $$A^{1/n} = \sqrt[n]{A}$$

4. $$A^{p/q} = (\sqrt[q]{A})^p = \sqrt[q]{A^p}$$

5. $$\sqrt[n]{A^{kn}} = A^{kn/n} = A^k$$

6. $$\sqrt[n]{AB} = \sqrt[n]{A}\sqrt[n]{B}$$

7. $$\frac{\sqrt[n]{A}}{\sqrt[n]{B}} = \sqrt[n]{\frac{A}{B}}$$

8. $$\sqrt[3]{\sqrt[n]{A}} = \sqrt[n]{\sqrt[3]{A}} = \sqrt[3n]{A}$$

Problems

1. Simplify the symbols by grouping and combining like terms.

   a. $(x + 3y - z) - (2y - x + 3z) + (4z - 3x + 2y)$

   Answer: $-x + 3\,y$

   b. $3(x^2 - 2yz + y^2) - 4(x^2 - y^2 - 3yz) + x^2 + y^2$

   Answer: $6yz + 8y^2$

   c. $3x + 4y + 3[x - 2(y - x) - y]$

   Answer: $12x - 5y$

   d. $3 - 2x - [-(x + y)] + [x - 2y]$

   Answer: $3 - y$

2. *Find the products:*

   a. $2xy(3x^2y - 4y^3)$

   Answer: $6x^3y^2 - 8xy^4$

   b. $3x^2y^3(2xy - x - 2y)$

   Answer: $6x^3y^4 - 3x^3y^3 - 6x^2y^4$

   c. $(3a + 5b)(3a - 5b)$

   Answer: $9a^2 - 25b^2$

   d. $(5xy + 4)(5xy - 4)$

   Answer: $25x^2y^2 - 16$

   e. $(2 - 5y^2)(2 + 5y^2)$

   Answer: $4 - 25y^4$

   f. $(2x + 1)^3$

   Answer: $8x^3 + 12x^2 + 6x + 1$

3. *Factor:*

   a. $3x^2y^4 + 6x^3y^3$

   Answer: $3x^2y^3\,(y + 2x)$

   b. $12s^2t^2 - 6s^5t^4 + 4s^4t$

   Answer: $2s^2t\,(6t - 3s^3t^3 + 2s^2)$

   c. $64x - x^3$

   Answer: $x\,(8 + x)(8 - x)$

   d. $18x^3y - 8xy^3$

   Answer: $2xy\,(3x + 2y)(3x - 2y)$

   e. $2z^3 + 10z^2 - 28z$

   Answer: $2z\,(z + 7)(z - 2)$

   f. $2s^4t - 4s^3t^2 - 6s^2t^3$

   Answer: $2s^2t\,(s - 3t)(s + t)$

4. *Factor and reduce:*

a.
$$\frac{24x^3y^2}{18xy^3}$$

Answer: $\dfrac{4x^2}{3y}$

b.
$$\frac{36xy^4z^2}{-15x^4y^3z}$$

Answer: $\dfrac{-12zy}{5x^3}$

c.
$$\frac{5a^2 - 10ab}{a - 2b}$$

Answer: 5a

d.
$$\frac{4x^2 - 16}{x^2 - 2x}$$

Answer: $\dfrac{4(x + 2)}{x}$

e.
$$\frac{3a^2}{4b^3}\left(\frac{2b^4}{9a^3}\right)$$

Answer: $\dfrac{b}{6a}$

f.
$$\frac{3x}{8y} \div \frac{9x}{16y}$$

Answer: $\dfrac{2}{3}$

g.
$$\frac{24x^3y^2}{5z^2} \div \frac{8x^2y^3}{15z^4}$$

Answer: $\dfrac{9xz^2}{y}$

h.
$$\frac{2x}{3} - \frac{x}{2}$$

Answer: $\dfrac{x}{6}$

i.
$$\frac{4}{3x} - \frac{5}{4x}$$

Answer: $\dfrac{1}{12x}$

5. *Evaluate:*

   a. $\qquad\qquad\qquad\qquad (-2x)^3$

Answer: $-8x^3$

   b. $\qquad\qquad\qquad\qquad \left(\dfrac{3y}{4}\right)^3$

Answer: $\dfrac{27y^3}{64}$

   c. $\qquad\qquad\qquad\qquad 4^{-3}$

Answer: $\dfrac{1}{64}$

   d. $\qquad\qquad\qquad\qquad (-4x)^{-2}$

Answer: $\dfrac{1}{16x^2}$

   e. $\qquad\qquad\qquad\qquad (2y^{-1})^{-1}$

Answer: $\dfrac{y}{2}$

   f. $\qquad\qquad\qquad\qquad \dfrac{3^{-1}x^2y^{-4}}{2^{-2}x^{-3}y^3}$

Answer: $\dfrac{4x^5}{3y^7}$

   g. $\qquad\qquad\qquad\qquad (16)^{1/4}$

Answer: 2

   h. $\qquad\qquad\qquad\qquad (-a^3b^3)^{-2/3}$

Answer: $\dfrac{1}{a^2b^2}$

   i. $\qquad\qquad\qquad\qquad (10^3)^0$

Answer: 1

   j. $\qquad\qquad\qquad\qquad 3y^{2/3}\,(y^{4/3})$

Answer: $3y^2$

   k. $\qquad\qquad\qquad\qquad \dfrac{1}{\sqrt[3]{8x}}\,(6)$

Answer: $\dfrac{3}{x^{1/3}}$

6. *Solve the following exponents:*

    a. $a^p(a^q)$

    Answer: a^{p+q}

    b. $a^3(a^5)$

    Answer: a^8

    c. $3^4(3^5)$

    Answer: 3^9

    d. $a^{n+1}(a^{n-2})$

    Answer: a^{2n-1}

    e. $[4(10^{-6})]\,[2(10^4)]$

    Answer: $8(10^{-2}) = 0.08$

    f. $\dfrac{a^p}{a^q}$

    Answer: a^{p-q}

    g. $\dfrac{a^5}{a^3}$

    Answer: a^2

    h. $(ab)^p$

    Answer: $a^p b^p$

    i. $(2a)^4$

    Answer: $2^4 a^4 = 16a^4$

    j. $\left(\dfrac{a}{b}\right)^p$

    Answer: $\dfrac{a^p}{b^p}$

    k. $\left(\dfrac{2}{3}\right)^4$

    Answer: $\dfrac{2^4}{3^4} = \dfrac{16}{81}$

    l. $\sqrt[3]{\dfrac{8x^{3n}}{27y^6}}$

    Answer: $\dfrac{2x^n}{3y^2}$

m.

$$\left(\frac{a^{1/3}}{x^{1/3}}\right)^{3/2}$$

Answer: $\dfrac{a^{1/2}}{x^{1/2}} = \dfrac{\sqrt{a}}{\sqrt{x}} = \sqrt{\dfrac{a}{x}}$

7. *Solve the following equations.*

   a. $3x - 2 = 7$

   Answer: $x = 3$

   b. $y + 3(y - 4) = 4$

   Answer: $y = 4$

   c. $4x - 3 = 5 - 2x$

   Answer: $x = 4/3$

   d. $\dfrac{2t - 9}{3} = \dfrac{3t + 4}{2}$

   Answer: $t = -6$

   e. $x^2 - 40 = 9$

   Answer: $x = \pm 7$

   f. $x^2 - 6 = x$

   Answer: $x = +3, -2$

   g. $x^2 - 7x = -12$

   Answer: $x = 3, 4$

   h. $x^2 + x = 6$

   Answer: $x = 2, -3$

   i. $x^2 = 5x + 24$

   Answer: $x = 8, -3$

   j. $9x^2 = 9x - 2$

   Answer: $x = 1/3, 2/3$

8. *Solve by quadratic formula:* $x = \dfrac{-b \pm \sqrt{b^2 - 4ac}}{2a}$

   a. $x^2 - 5x = 6$

   Answer: $x = 6, -1$

b. $$x^2 - 6 = x$$

Answer: 3, -2

c. $$3x^2 - 2x = 8$$

Answer: $x = 2, -4/3$

9. *Express each as an algebraic sum of logarithms:*

    a. $$\log_b UVW$$

    Answer: $\log_b U + \log_b V + \log_b W$

    b. $$\log_b \frac{UV}{W}$$

    Answer: $\log_b U + \log_b V - \log_b W$

    c. $$\log_b \frac{XYZ}{PQ}$$

    Answer: $\log_b X + \log_b Y + \log_b Z - (\log_b P + \log_b Q)$

    d. $$\log_b \frac{U^2}{V^3}$$

    Answer: $2 \log_b U - 3 \log_b V$

    e. $$\log_b \frac{U^2 V^3}{W^4}$$

    Answer: $2 \log_b U + 3 \log_b V - 4 \log_b W$

    f. $$\log_e \frac{\sqrt{x^3}}{\sqrt[4]{y^3}}$$

    Answer: $\frac{3}{2} \log_e X - \frac{3}{4} \log_e Y$ *or* $\frac{3}{2} \ln X - \frac{3}{4} \ln Y$

10. *Graph*

    a. $$y = x - 2$$

    Answer:

x	y
0	-2
1	-1
2	0
-1	-3
-2	-4

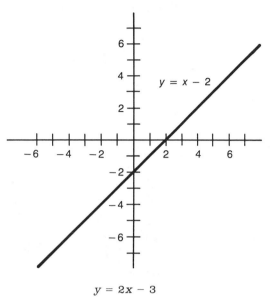

$$y = x - 2$$

b. $\qquad y = 2x - 3$

Answer:

x	y
0	−3
1	−1
2	1
−1	−5

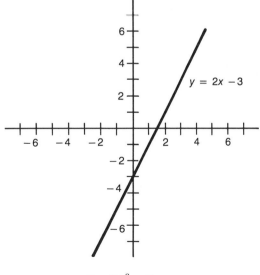

$$y = 2x - 3$$

c. $\qquad y = 2x^2 - 2$

Answer:

x	y
0	−2
1	0
2	6
−1	0
−2	6

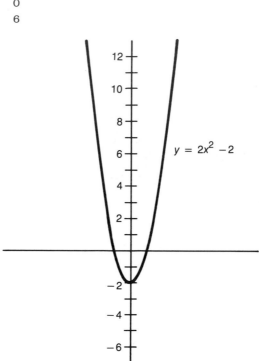

$y = 2x^2 - 2$

Table A1. Areas under the Normal Curve[a]
(Proportion of total area under the curve from − ∞ to designated Z value)

Z	0.09	0.08	0.07	0.06	0.05	0.04	0.03	0.02	0.01	0.00
−3.5	0.00017	0.00017	0.00018	0.00019	0.00019	0.00020	0.00021	0.00022	0.00022	0.00023
−3.4	0.00024	0.00025	0.00026	0.00027	0.00028	0.00029	0.00030	0.00031	0.00033	0.00034
−3.3	0.00035	0.00036	0.00038	0.00039	0.00040	0.00042	0.00043	0.00045	0.00047	0.00048
−3.2	0.00050	0.00052	0.00054	0.00056	0.00058	0.00060	0.00062	0.00064	0.00066	0.00069
−3.1	0.00071	0.00074	0.00076	0.00079	0.00082	0.00085	0.00087	0.00090	0.00094	0.00097
−3.0	0.00100	0.00104	0.00107	0.00111	0.00114	0.00118	0.00122	0.00126	0.00131	0.00135
−2.9	0.0014	0.0014	0.0015	0.0015	0.0016	0.0016	0.0017	0.0017	0.0018	0.0019
−2.8	0.0019	0.0020	0.0021	0.0021	0.0022	0.0023	0.0023	0.0024	0.0025	0.0026
−2.7	0.0026	0.0027	0.0028	0.0029	0.0030	0.0031	0.0032	0.0033	0.0034	0.0035
−2.6	0.0036	0.0037	0.0038	0.0039	0.0040	0.0041	0.0043	0.0044	0.0045	0.0047
−2.5	0.0048	0.0049	0.0051	0.0052	0.0054	0.0055	0.0057	0.0059	0.0060	0.0062
−2.4	0.0064	0.0066	0.0068	0.0069	0.0071	0.0073	0.0075	0.0078	0.0080	0.0082
−2.3	0.0084	0.0087	0.0089	0.0091	0.0094	0.0096	0.0099	0.0102	0.0104	0.0107
−2.2	0.0110	0.0113	0.0116	0.0119	0.0122	0.0125	0.0129	0.0132	0.0136	0.0139
−2.1	0.0143	0.0146	0.0150	0.0154	0.0158	0.0162	0.0166	0.0170	0.0174	0.0179
−2.0	0.0183	0.0188	0.0192	0.0197	0.0202	0.0207	0.0212	0.0217	0.0222	0.0228
−1.9	0.0233	0.0239	0.0244	0.0250	0.0256	0.0262	0.0268	0.0274	0.0281	0.0287
−1.8	0.0294	0.0301	0.0307	0.0314	0.0322	0.0329	0.0336	0.0344	0.0351	0.0359
−1.7	0.0367	0.0375	0.0384	0.0392	0.0401	0.0409	0.0418	0.0427	0.0436	0.0446
−1.6	0.0455	0.0465	0.0475	0.0485	0.0495	0.0505	0.0516	0.0526	0.0537	0.0548
−1.5	0.0559	0.0571	0.0582	0.0594	0.0606	0.0618	0.0630	0.0643	0.0655	0.0668
−1.4	0.0681	0.0694	0.0708	0.0721	0.0735	0.0749	0.0764	0.0778	0.0793	0.0808
−1.3	0.0823	0.0838	0.0853	0.0869	0.0885	0.0901	0.0918	0.0934	0.0951	0.0968
−1.2	0.0985	0.1003	0.1020	0.1038	0.1057	0.1075	0.1093	0.1112	0.1131	0.1151
−1.1	0.1170	0.1190	0.1210	0.1230	0.1251	0.1271	0.1292	0.1314	0.1335	0.1357
−1.0	0.1379	0.1401	0.1423	0.1446	0.1469	0.1492	0.1515	0.1539	0.1562	0.1587
−0.9	0.1611	0.1635	0.1660	0.1685	0.1711	0.1736	0.1762	0.1788	0.1814	0.1841
−0.8	0.1867	0.1894	0.1922	0.1949	0.1977	0.2005	0.2033	0.2061	0.2090	0.2119
−0.7	0.2148	0.2177	0.2207	0.2236	0.2266	0.2297	0.2327	0.2358	0.2389	0.2420
−0.6	0.2451	0.2483	0.2514	0.2546	0.2578	0.2611	0.2643	0.2676	0.2709	0.2743
−0.5	0.2776	0.2810	0.2843	0.2877	0.2912	0.2946	0.2981	0.3015	0.3050	0.3085
−0.4	0.3121	0.3156	0.3192	0.3228	0.3264	0.3300	0.3336	0.3372	0.3409	0.3446
−0.3	0.3483	0.3520	0.3557	0.3594	0.3632	0.3669	0.3707	0.3745	0.3783	0.3821
−0.2	0.3859	0.3897	0.3936	0.3974	0.4013	0.4052	0.4090	0.4129	0.4168	0.4207
−0.1	0.4247	0.4286	0.4325	0.4364	0.4404	0.4443	0.4483	0.4522	0.4562	0.4602
−0.0	0.4641	0.4681	0.4721	0.4761	0.4801	0.4840	0.4880	0.4920	0.4960	0.5000

Table A1. Areas under the Normal Curve *(Continued)*

Z	0.00	0.01	0.02	0.03	0.04	0.05	0.06	0.07	0.08	0.09
+0.0	0.5000	0.5040	0.5080	0.5120	0.5160	0.5199	0.5239	0.5279	0.5319	0.5359
+0.1	0.5398	0.5438	0.5478	0.5517	0.5557	0.5596	0.5636	0.5675	0.5714	0.5753
+0.2	0.5793	0.5832	0.5871	0.5910	0.5948	0.5987	0.6026	0.6064	0.6103	0.6141
+0.3	0.6179	0.6217	0.6255	0.6293	0.6331	0.6368	0.6406	0.6443	0.6480	0.6517
+0.4	0.6554	0.6591	0.6628	0.6664	0.6700	0.6736	0.6772	0.6808	0.6844	0.6879
+0.5	0.6915	0.6950	0.6985	0.7019	0.7054	0.7088	0.7123	0.7157	0.7190	0.7224
+0.6	0.7257	0.7291	0.7324	0.7357	0.7389	0.7422	0.7454	0.7486	0.7517	0.7549
+0.7	0.7580	0.7611	0.7642	0.7673	0.7704	0.7734	0.7764	0.7794	0.7823	0.7852
+0.8	0.7881	0.7910	0.7939	0.7967	0.7995	0.8023	0.8051	0.8079	0.8106	0.8133
+0.9	0.8159	0.8186	0.8212	0.8238	0.8264	0.8289	0.8315	0.8340	0.8365	0.8389
+1.0	0.8413	0.8438	0.8461	0.8485	0.8508	0.8531	0.8554	0.8577	0.8599	0.8621
+1.1	0.8643	0.8665	0.8686	0.8708	0.8729	0.8749	0.8770	0.8790	0.8810	0.8830
+1.2	0.8849	0.8869	0.8888	0.8907	0.8925	0.8944	0.8962	0.8980	0.8997	0.9015
+1.3	0.9032	0.9049	0.9066	0.9082	0.9099	0.9115	0.9131	0.9147	0.9162	0.9177
+1.4	0.9192	0.9207	0.9222	0.9236	0.9251	0.9265	0.9279	0.9292	0.9306	0.9319
+1.5	0.9332	0.9345	0.9357	0.9370	0.9382	0.9394	0.9406	0.9418	0.9429	0.9441
+1.6	0.9452	0.9463	0.9474	0.9484	0.9495	0.9505	0.9515	0.9525	0.9535	0.9545
+1.7	0.9554	0.9564	0.9573	0.9582	0.9591	0.9599	0.9608	0.9616	0.9625	0.9633
+1.8	0.9641	0.9649	0.9656	0.9664	0.9671	0.9678	0.9686	0.9693	0.9699	0.9706
+1.9	0.9713	0.9719	0.9726	0.9732	0.9738	0.9744	0.9750	0.9756	0.9761	0.9767
+2.0	0.9773	0.9778	0.9783	0.9788	0.9793	0.9798	0.9803	0.9808	0.9812	0.9817
+2.1	0.9821	0.9826	0.9830	0.9834	0.9838	0.9842	0.9846	0.9850	0.9854	0.9857
+2.2	0.9861	0.9864	0.9868	0.9871	0.9875	0.9878	0.9881	0.9884	0.9887	0.9890
+2.3	0.9893	0.9896	0.9898	0.9901	0.9904	0.9906	0.9909	0.9911	0.9913	0.9916
+2.4	0.9918	0.9920	0.9922	0.9925	0.9927	0.9929	0.9931	0.9932	0.9934	0.9936
+2.5	0.9938	0.9940	0.9941	0.9943	0.9945	0.9946	0.9948	0.9949	0.9951	0.9952
+2.6	0.9953	0.9955	0.9956	0.9957	0.9959	0.9960	0.9961	0.9962	0.9963	0.9964
+2.7	0.9965	0.9966	0.9967	0.9968	0.9969	0.9970	0.9971	0.9972	0.9973	0.9974
+2.8	0.9974	0.9975	0.9976	0.9977	0.9977	0.9978	0.9979	0.9979	0.9980	0.9981
+2.9	0.9981	0.9982	0.9983	0.9983	0.9984	0.9984	0.9985	0.9985	0.9986	0.9986
+3.0	0.99865	0.99869	0.99874	0.99878	0.99882	0.99886	0.99889	0.99893	0.99896	0.99900
+3.1	0.99903	0.99906	0.99910	0.99913	0.99915	0.99918	0.99921	0.99924	0.99926	0.99929
+3.2	0.99931	0.99934	0.99936	0.99938	0.99940	0.99942	0.99944	0.99946	0.99948	0.99950
+3.3	0.99952	0.99953	0.99955	0.99957	0.99958	0.99960	0.99961	0.99962	0.99964	0.99965
+3.4	0.99966	0.99967	0.99969	0.99970	0.99971	0.99972	0.99973	0.99974	0.99975	0.99976
+3.5	0.99977	0.99978	0.99978	0.99979	0.99980	0.99981	0.99981	0.99982	0.99983	0.99983

[a]By permission of McGraw-Hill Book Co.

$z = (x_i - \mu)/\sigma$

$P_z = P(z \leq z_i)$

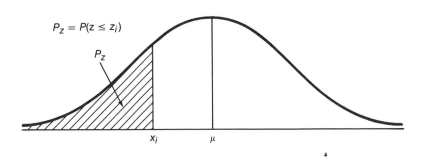

$P_z = P(z \leq z_i)$

P_z

x_i μ

Table A2. Summation of Terms of Poisson's Exponential Binomial Limit[a]

(1000 × Probability of c or Less Occurrences of Event That Has Average Number of Occurrences Equal to np or λT)

np / λT	c 0	1	2	3	4	5	6	7	8	9
0.02	980	1,000								
0.04	961	999	1,000							
0.06	942	998	1,000							
0.08	923	997	1,000							
0.10	905	995	1,000							
0.15	861	990	999	1,000						
0.20	819	982	999	1,000						
0.25	779	974	998	1,000						
0.30	741	963	996	1,000						
0.35	705	951	994	1,000						
0.40	670	938	992	999	1,000					
0.45	638	925	989	999	1,000					
0.50	607	910	986	998	1,000					
0.55	577	894	982	998	1,000					
0.60	549	878	977	997	1,000					
0.65	522	861	972	996	999	1,000				
0.70	497	844	966	994	999	1,000				
0.75	472	827	959	993	999	1,000				
0.80	449	809	953	991	999	1,000				
0.85	427	791	945	989	998	1,000				
0.90	407	772	937	987	998	1,000				
0.95	387	754	929	984	997	1,000				
1.00	368	736	920	981	996	999	1,000			
1.1	333	699	900	974	995	999	1,000			
1.2	301	663	879	966	992	998	1,000			
1.3	273	627	857	957	989	998	1,000			
1.4	247	592	833	946	986	997	999	1,000		
1.5	223	558	809	934	981	996	999	1,000		
1.6	202	525	783	921	976	994	999	1,000		
1.7	183	493	757	907	970	992	998	1,000		
1.8	165	463	731	891	964	990	997	999	1,000	
1.9	150	434	704	875	956	987	997	999	1,000	
2.0	135	406	677	857	947	983	995	999	1,000	

[a]By permission of McGraw-Hill Book Co.

Table A2. Summation of Terms of Poisson's Exponential Binomial Limit *(Continued)*

np / λT	0	1	2	3	4	5	6	7	8	9
2.2	111	355	623	819	928	975	993	998	1,000	
2.4	091	308	570	779	904	964	988	997	999	1,000
2.6	074	267	518	736	877	951	983	995	999	1,000
2.8	061	231	469	692	848	935	976	992	998	999
3.0	050	199	423	647	815	916	966	988	996	999
3.2	041	171	380	603	781	895	955	983	994	998
3.4	033	147	340	558	744	871	942	977	992	997
3.6	027	126	303	515	706	844	927	969	988	996
3.8	022	107	269	473	668	816	909	960	984	994
4.0	018	092	238	433	629	785	889	949	979	992
4.2	015	078	210	395	590	753	867	936	972	989
4.4	012	066	185	359	551	720	844	921	964	985
4.6	010	056	163	326	513	686	818	905	955	980
4.8	008	048	143	294	476	651	791	887	944	975
5.0	007	040	125	265	440	616	762	867	932	968
5.2	006	034	109	238	406	581	732	845	918	960
5.4	005	029	095	213	373	546	702	822	903	951
5.6	004	024	082	191	342	512	670	797	886	941
5.8	003	021	072	170	313	478	638	771	867	929
6.0	002	017	062	151	285	446	606	744	847	916

	10	11	12	13	14	15	16
2.8	1,000						
3.0	1,000						
3.2	1,000						
3.4	999	1,000					
3.6	999	1,000					
3.8	998	999	1,000				
4.0	997	999	1,000				
4.2	996	999	1,000				
4.4	994	998	999	1,000			
4.6	992	997	999	1,000			
4.8	990	996	999	1,000			
5.0	986	995	998	999	1,000		
5.2	982	993	997	999	1,000		
5.4	977	990	996	999	1,000		
5.6	972	988	995	998	999	1,000	
5.8	965	984	993	997	999	1,000	
6.0	957	980	991	996	999	999	1,000

Table A2. Summation of Terms of Poisson's Exponential Binomial Limit *(Continued)*

np λT \ c	0	1	2	3	4	5	6	7	8	9
6.2	002	015	054	134	259	414	574	716	826	902
6.4	002	012	046	119	235	384	542	687	803	886
6.6	001	010	040	105	213	355	511	658	780	869
6.8	001	009	034	093	192	327	480	628	755	850
7.0	001	007	030	082	173	301	450	599	729	830
7.2	001	006	025	072	156	276	420	569	703	810
7.4	001	005	022	063	140	253	392	539	676	788
7.6	001	004	019	055	125	231	365	510	648	765
7.8	000	004	016	048	112	210	338	481	620	741
8.0	000	003	014	042	100	191	313	453	593	717
8.5	000	002	009	030	074	150	256	386	523	653
9.0	000	001	006	021	055	116	207	324	456	587
9.5	000	001	004	015	040	089	165	269	392	522
10.0	000	000	003	010	029	067	130	220	333	458

	10	11	12	13	14	15	16	17	18	19
6.2	949	975	989	995	998	999	1,000			
6.4	939	969	986	994	997	999	1,000			
6.6	927	963	982	992	997	999	999	1,000		
6.8	915	955	978	990	996	998	999	1,000		
7.0	901	947	973	987	994	998	999	1,000		
7.2	887	937	967	984	993	997	999	999	1,000	
7.4	871	926	961	980	991	996	998	999	1,000	
7.6	854	915	954	976	989	995	998	999	1,000	
7.8	835	902	945	971	986	993	997	999	1,000	
8.0	816	888	936	966	983	992	996	998	999	1,000
8.5	763	849	909	949	973	986	993	997	999	999
9.0	706	803	876	926	959	978	989	995	998	999
9.5	645	752	836	898	940	967	982	991	996	998
10.0	583	697	792	864	917	951	973	986	993	997

	20	21	22
8.5	1,000		
9.0	1,000		
9.5	999	1,000	
10.0	998	999	1,000

Table A2. Summation of Terms of Poisson's Exponential Binomial Limit *(Continued)*

np / λT \ c	0	1	2	3	4	5	6	7	8	9
10.5	000	000	002	007	021	050	102	179	279	397
11.0	000	000	001	005	015	038	079	143	232	341
11.5	000	000	001	003	011	028	060	114	191	289
12.0	000	000	001	002	008	020	046	090	155	242
12.5	000	000	000	002	005	015	035	070	125	201
13.0	000	000	000	001	004	011	026	054	100	166
13.5	000	000	000	001	003	008	019	041	079	135
14.0	000	000	000	000	002	006	014	032	062	109
14.5	000	000	000	000	001	004	010	024	048	088
15.0	000	000	000	000	001	003	008	018	037	070

	10	11	12	13	14	15	16	17	18	19
10.5	521	639	742	825	888	932	960	978	988	994
11.0	460	579	689	781	854	907	944	968	982	991
11.5	402	520	633	733	815	878	924	954	974	986
12.0	347	462	576	682	772	844	899	937	963	979
12.5	297	406	519	628	725	806	869	916	948	969
13.0	252	353	463	573	675	764	835	890	930	957
13.5	211	304	409	518	623	718	798	861	908	942
14.0	176	260	358	464	570	669	756	827	883	923
14.5	145	220	311	413	518	619	711	790	853	901
15.0	118	185	268	363	466	568	664	749	819	875

	20	21	22	23	24	25	26	27	28	29
10.5	997	999	999	1,000						
11.0	995	998	999	1,000						
11.5	992	996	998	999	1,000					
12.0	988	994	997	999	999	1,000				
12.5	983	991	995	998	999	999	1,000			
13.0	975	986	992	996	998	999	1,000			
13.5	965	980	989	994	997	998	999	1,000		
14.0	952	971	983	991	995	997	999	999	1,000	
14.5	936	960	976	986	992	996	998	999	999	1,000
15.0	917	947	967	981	989	994	997	998	999	1,000

Table A2. Summation of Terms of Poisson's Exponential Binomial Limit (Continued)

np λT \ c	4	5	6	7	8	9	10	11	12	13
16	000	001	004	010	022	043	077	127	193	275
17	000	001	002	005	013	026	049	085	135	201
18	000	000	001	003	007	015	030	055	092	143
19	000	000	001	002	004	009	018	035	061	098
20	000	000	000	001	002	005	011	021	039	066
21	000	000	000	000	001	003	006	013	025	043
22	000	000	000	000	001	002	004	008	015	028
23	000	000	000	000	000	001	002	004	009	017
24	000	000	000	000	000	000	001	003	005	011
25	000	000	000	000	000	000	001	001	003	006

	14	15	16	17	18	19	20	21	22	23
16	368	467	566	659	742	812	868	911	942	963
17	281	371	468	564	655	736	805	861	905	937
18	208	287	375	469	562	651	731	799	855	899
19	150	215	292	378	469	561	647	725	793	849
20	105	157	221	297	381	470	559	644	721	787
21	072	111	163	227	302	384	471	558	640	716
22	048	077	117	169	232	306	387	472	556	637
23	031	052	082	123	175	238	310	389	472	555
24	020	034	056	087	128	180	243	314	392	473
25	012	022	038	060	092	134	185	247	318	394

	24	25	26	27	28	29	30	31	32	33
16	978	987	993	996	998	999	999	1,000		
17	959	975	985	991	995	997	999	999	1,000	
18	932	955	972	983	990	994	997	998	999	1,000
19	893	927	951	969	980	988	993	996	998	999
20	843	888	922	948	966	978	987	992	995	997
21	782	838	883	917	944	963	976	985	991	994
22	712	777	832	877	913	940	959	973	983	989
23	635	708	772	827	873	908	936	956	971	981
24	554	632	704	768	823	868	904	932	953	969
25	473	553	629	700	763	818	863	900	929	950

	34	35	36	37	38	39	40	41	42	43
19	999	1,000								
20	999	999	1,000							
21	997	998	999	999	1,000					
22	994	996	998	999	999	1,000				
23	988	993	996	997	999	999	1,000			
24	979	987	992	995	997	998	999	999	1,000	
25	966	978	985	991	994	997	998	999	999	1,000

Table A3. Table of Chi Square[a]

(For larger values of ν, the expression $\sqrt{2\chi^2} - \sqrt{2\nu - 1}$ may be used as a normal deviate with unit variance, remembering that the probability for χ^2 corresponds with that of a single tail of the normal curve; $\nu =$ degrees of freedom $= 2f$)

2f	Probability (α)										
ν	0.99	0.98	0.95	0.90	0.80	0.20	0.10	0.05	0.02	0.01	0.001
1	0.000157	0.000628	0.00393	0.0158	0.0642	1.642	2.706	3.841	5.412	6.635	10.827
2	0.0201	0.0404	0.103	0.211	0.446	3.219	4.605	5.991	7.824	9.210	13.815
3	0.115	0.185	0.352	0.584	1.005	4.642	6.251	7.815	9.837	11.341	16.268
4	0.297	0.429	0.711	1.064	1.649	5.989	7.779	9.488	11.668	13.277	18.465
5	0.554	0.752	1.145	1.610	2.343	7.289	9.236	11.070	13.388	15.086	20.517
6	0.872	1.134	1.635	2.204	3.070	8.558	10.645	12.592	15.033	16.812	22.457
7	1.239	1.564	2.167	2.833	3.822	9.803	12.017	14.067	16.622	18.475	24.322
8	1.646	2.032	2.733	3.490	4.594	11.030	13.362	15.507	18.168	20.090	26.125
9	2.088	2.532	3.325	4.168	5.380	12.242	14.684	16.919	19.679	21.666	27.877
10	2.558	3.059	3.940	4.865	6.179	13.442	15.987	18.307	21.161	23.209	29.588
11	3.053	3.609	4.575	5.578	6.989	14.631	17.275	19.675	22.618	24.725	31.264
12	3.571	4.178	5.226	6.304	7.807	15.812	18.549	21.026	24.054	26.217	32.909
13	4.107	4.765	5.892	7.042	8.634	16.985	19.812	22.362	25.472	27.688	34.528
14	4.660	5.368	6.571	7.790	9.467	18.151	21.064	23.685	26.873	29.141	36.123
15	5.229	5.985	7.261	8.547	10.307	19.311	22.307	24.996	28.259	30.578	37.697
16	5.812	6.614	7.962	9.312	11.152	20.465	23.542	26.296	29.633	32.000	39.252
17	6.408	7.255	8.672	10.085	12.002	21.615	24.769	27.587	30.995	33.409	40.790
18	7.015	7.906	9.390	10.865	12.857	22.760	25.989	28.869	32.346	34.805	42.312
19	7.633	8.567	10.117	11.651	13.716	23.900	27.204	30.144	33.687	36.191	43.820
20	8.260	9.237	10.851	12.443	14.578	25.038	28.412	31.410	35.020	37.566	45.315
21	8.897	9.915	11.591	13.240	15.445	26.171	29.615	32.671	36.343	38.932	46.797
22	9.542	10.600	12.338	14.041	16.314	27.301	30.813	33.924	37.659	40.289	48.268
23	10.196	11.293	13.091	14.848	17.187	28.429	32.007	35.172	38.968	41.638	49.728
24	10.856	11.992	13.848	15.659	18.062	29.553	33.196	36.415	40.270	42.980	51.179
25	11.524	12.697	14.611	16.473	18.940	30.675	34.382	37.652	41.566	44.314	52.620
26	12.198	13.409	15.379	17.292	19.820	31.795	35.563	38.885	42.856	45.642	54.052
27	12.879	14.125	16.151	18.114	20.703	32.912	36.741	40.113	44.140	46.963	55.476
28	13.565	14.847	16.928	18.939	21.588	34.027	37.916	41.337	45.419	48.278	56.893
29	14.256	15.574	17.708	19.768	22.475	35.139	39.087	42.557	46.693	49.588	58.302
30	14.953	16.306	18.493	20.599	23.364	36.250	40.256	43.773	47.962	50.892	59.703

[a] By permission of McGraw-Hill Book Co.

Table A4. Percentiles of the t Distribution[a]

df	$t_{0.60}$	$t_{0.70}$	$t_{0.80}$	$t_{0.90}$	$t_{0.95}$	$t_{0.975}$	$t_{0.99}$	$t_{0.995}$
1	0.325	0.727	1.376	3.078	6.314	12.706	31.821	63.657
2	0.280	0.617	1.061	1.886	2.920	4.303	6.965	9.925
3	0.277	0.584	0.978	1.638	2.353	3.182	4.541	5.841
4	0.271	0.569	0.941	1.533	2.132	2.776	3.747	4.604
5	0.267	0.559	0.920	1.476	2.015	2.571	3.365	4.032
6	0.265	0.553	0.906	1.440	1.943	2.447	3.143	3.707
7	0.263	0.549	0.896	1.415	1.895	2.365	2.998	3.499
8	0.262	0.546	0.889	1.397	1.860	2.306	2.896	3.355
9	0.261	0.543	0.883	1.383	1.833	2.262	2.821	3.250
10	0.260	0.542	0.879	1.372	1.812	2.228	2.764	3.169
11	0.260	0.540	0.876	1.363	1.796	2.201	2.718	3.106
12	0.259	0.539	0.873	1.356	1,782	2.179	2.681	3.055
13	0.259	0.538	0.870	1.350	1.771	2.160	2.650	3.012
14	0.258	0.537	0.868	1.345	1,761	2.145	2.624	2.977
15	0.258	0.536	0.866	1.341	1.753	2.131	2.602	2.947
16	0.258	0.535	0.865	1.337	1.746	2.120	2.583	2.921
17	0.257	0.534	0.863	1.333	1.740	2.110	2.567	2.898
18	0.257	0.534	0.862	1.330	1.734	2.101	2.552	2.878
19	0.257	0.533	0.861	1.328	1.729	2.093	2.539	2.861
20	0.257	0.533	0.860	1.325	1.725	2.086	2.528	2.845
21	0.257	0.532	0.859	1.323	1.721	2.080	2.518	2.831
22	0.256	0.532	0.858	1.321	1.717	2.074	2.508	2.819
23	0.256	0.532	0.858	1.319	1.714	2.069	2.500	2.807
24	0.256	0.531	0.857	1.318	1.711	2.064	2.492	2.797
25	0.256	0.531	0.856	1.316	1.708	2.060	2.485	2.787
26	0.256	0.531	0.856	1.315	1.706	2.056	2.479	2.779
27	0.256	0.531	0.855	1.314	1.703	2.052	2.473	2.771
28	0.256	0.530	0.855	1.313	1.701	2.048	2.467	2.763
29	0.256	0.530	0.854	1.311	1.699	2.045	2.462	2.756
30	0.256	0.530	0.854	1.310	1.697	2.042	2.457	2.750
40	0.255	0.529	0.851	1.303	1.684	2.021	2.423	2.704
60	0.254	0.527	0.848	1.296	1.671	2.000	2.390	2.660
120	0.254	0.526	0.845	1.289	1.658	1.980	2.358	2.617
∞	0.253	0.524	0.842	1.282	1.645	1.960	2.326	2.576
df	$-t_{0.40}$	$-t_{0.30}$	$-t_{0.20}$	$-t_{0.10}$	$-t_{0.05}$	$-t_{0.025}$	$-t_{0.01}$	$-t_{0.005}$

When the table is read from the foot, the tabled values are to be prefixed with a negative sign. Interpolation should be performed using the reciprocals of the degrees of freedom.

[a] By permission of McGraw-Hill Book Co.

Table A5. Weibull Ranking Percentiles
(For sample sizes of 20 or less)[a]

Confidence Level =:										
0.5	1.0	2.5	5.0	10.0	50.0	90.0	95.0	97.5	99.0	99.5
					Sample Size = 5					
0.1	0.2	0.5	1.0	2.1	13	37	45	52	60	65
2.3	3.3	5.3	8	11	31	58	66	72	78	81
8	11	15	19	25	50	75	81	85	89	92
19	22	28	34	42	69	89	92	95	97	98
35	40	48	55	63	87	98	99	99.5	99.8	99.9
					Sample Size = 6					
0.1	0.2	0.4	0.9	1.8	11	32	39	46	54	59
1.9	2.7	4.3	6.3	9	26	51	58	64	71	75
7	8	12	15	20	42	67	73	78	83	86
14	17	22	27	33	58	80	85	88	92	93
25	29	36	42	49	74	91	94	96	97	98
41	46	54	61	68	89	98	99	99.6	99.8	99.9
					Sample Size = 7					
0.07	0.1	0.4	0.7	1.5	9	28	35	41	48	53
1.6	2.3	3.7	5.3	7.9	22	45	52	58	64	68
5.5	7	10	13	17	36	60	66	71	76	80
12	14	18	23	28	50	72	77	82	86	88
20	24	29	34	40	64	83	87	90	93	94
32	36	42	48	55	77	92	95	96	97.7	98.4
47	52	59	65	72	91	98.5	99.3	99.6	99.8	99.9
					Sample Size = 8					
0.06	0.1	0.3	0.6	1.3	8	25	31	37	44	48
1.4	2.0	3.2	4.6	7	20	41	47	53	59	63
4.7	6	8	11	15	32	54	60	65	71	74
10	12	16	19	24	44	66	71	76	80	83
17	20	24	29	34	56	76	81	84	88	90
26	29	35	40	46	68	85	89	91	94	95
37	41	47	53	59	80	93	95	97	98	99
52	56	63	69	75	92	98.7	99.4	99.7	99.8	99.9
					Sample Size = 9					
0.06	0.1	0.3	0.7	1.2	7	23	28	34	40	44
1.2	1.7	2.8	4.1	6	18	37	43	48	54	58
4.2	5.3	7	10	13	29	49	55	60	66	69
9	11	14	17	21	39	60	66	70	75	78
15	17	21	25	30	50	70	75	79	83	85
22	25	30	34	40	61	79	83	86	89	91
31	34	40	45	51	71	87	90	93	95	96
42	46	52	57	63	82	94	96	97	98	99
56	60	66	72	77	93	98.8	99.4	99.7	99.8	99.9

Table A5. Weibull Ranking Percentiles (*continued*)
(For sample sizes of 20 or less)[a]

Confidence Level =										
0.5	1.0	2.5	5.0	10.0	50.0	90.0	95.0	97.5	99.0	99.5
					Sample Size = 10					
0.05	0.1	0.3	0.5	1.0	7	21	26	31	37	41
1.1	1.6	2.5	3.7	5.5	16	36	39	45	50	54
3.7	4.7	6.7	9	12	26	45	51	56	61	65
8	9	12	15	19	36	55	61	65	70	74
13	15	19	22	27	45	65	70	74	78	81
19	22	26	30	35	55	73	78	81	85	87
26	30	35	39	45	64	81	85	88	91	92
35	39	44	49	55	74	88	91	93	95	96
46	50	55	61	66	84	95	96	97	98	99
59	63	69	74	79	93	99	99.5	99.7	99.9	99.9
					Sample Size = 11					
0.05	0.1	0.2	0.5	1.0	6	19	24	28	34	38
1.0	1.4	2.3	3.3	5	15	31	36	41	47	50
3.3	4.3	6	8	10	24	42	47	52	57	61
7	8	11	14	17	32	51	56	61	66	69
11	13	17	20	24	41	60	65	69	74	77
17	19	23	27	32	50	68	73	77	81	83
23	26	31	35	40	59	76	80	83	87	89
31	34	39	44	49	68	83	86	89	92	93
39	43	48	53	58	76	90	92	94	96	97
49	53	59	64	69	85	95	97	98	98.6	99
62	66	72	76	81	94	99	99.5	99.8	99.9	99.9
					Sample Size = 12					
0.04	0.1	0.2	0.4	0.9	6	17	22	26	32	36
0.9	1.3	2.1	3.0	4.5	14	29	34	38	44	48
3	4	5	7	10	22	39	44	48	54	57
6	8	10	12	15	30	48	53	57	62	66
10	12	15	18	22	38	56	61	65	70	73
15	17	21	25	29	46	64	68	72	77	79
21	23	28	32	36	54	71	75	79	83	85
27	30	35	39	44	62	78	82	85	88	89
34	38	43	47	52	70	85	88	90	92	94
43	46	52	56	61	78	90	93	95	96	97
53	56	62	66	71	86	95	97	98	98.7	99.1
64	68	74	78	83	94	99.1	99.6	99.8	99.9	99.9

Table A5. Weibull Ranking Percentiles (*continued*)
(For sample sizes of 20 or less)[a]

Confidence Level =										
0.5	1.0	2.5	5.0	10.0	50.0	90.0	95.0	97.5	99.0	99.5
Sample Size = 13										
0.04	0.08	0.2	0.4	0.8	5	16	21	25	30	33
0.8	1.2	1.9	2.8	4.2	13	27	32	36	41	45
2.8	3.6	5	7	9	20	36	41	45	51	54
6	7	9	11	14	28	44	49	54	59	62
9	11	14	17	20	35	52	57	61	66	69
14	16	19	22	26	43	60	85	68	73	75
19	21	25	29	33	50	67	71	75	79	81
25	27	32	35	40	57	74	78	81	84	86
31	34	39	43	48	65	80	83	86	89	91
38	41	46	51	56	72	86	89	91	93	94
46	49	55	59	64	80	91	93	95	96	97
55	59	64	68	73	87	96	97	98	98.8	99.2
67	70	75	79	83	95	99	99.6	99.8	99.9	99.9
Sample Size = 14										
0.04	0.07	0.2	0.4	0.8	5	15	19	23	28	32
0.8	1.1	1.8	2.6	3.9	12	25	30	34	39	42
2.6	3.3	4.7	6	8	19	34	39	43	48	51
5	6	8	10	13	26	42	47	51	56	59
9	10	13	15	19	33	49	54	58	63	66
13	15	18	21	24	40	56	61	65	69	72
17	19	23	26	30	47	63	67	71	75	78
22	25	29	33	37	53	70	74	77	81	83
28	31	35	39	44	60	76	79	82	85	87
34	37	42	46	51	67	81	85	87	90	91
41	44	49	53	58	74	87	90	92	94	95
49	52	57	61	66	81	92	94	95	96	97
58	61	66	70	75	88	96	97	98	99	99.2
68	72	77	81	85	95	99	99.6	99.8	99.9	99.9
Sample Size = 15										
0.03	0.07	0.2	0.3	0.7	4.5	14	18	22	26	30
0.7	1.0	1.7	2.4	3.6	11	24	28	32	37	40
2.4	3.1	4.3	5.7	7.6	17	32	36	40	45	49
4.9	6	8	10	12	24	39	44	48	53	56
8	9	12	14	17	30	46	51	55	60	63
12	13	16	19	23	37	53	58	62	66	69
16	18	21	24	28	43	60	64	68	72	74
21	23	27	30	34	50	66	70	73	77	79
26	28	32	36	40	57	72	76	79	82	84
31	34	38	42	47	63	77	81	84	87	88
32	40	45	49	54	70	83	86	88	91	92
44	47	52	56	61	76	88	90	92	94	95
51	55	60	64	68	83	92	94	96	97	98
60	63	68	72	76	89	96	97	98	99	99.3
70	74	78	82	86	95	99	99.7	99.8	99.9	99.9

Table A5. Weibull Ranking Percentiles (continued)
(For sample sizes of 20 or less)[a]

					Confidence Level =					
0.5	1.0	2.5	5.0	10.0	50.0	90.0	95.0	97.5	99.0	99.5
					Sample Size = 16					
0.03	0.06	0.1	0.3	0.7	4	13	17	21	25	28
0.7	1.0	1.6	2.3	3.4	10	22	26	30	35	38
2.2	2.9	4.0	5.3	7	16	30	34	38	43	46
4.5	5.5	7	9	11	22	37	42	46	50	53
7	9	11	13	16	29	44	48	52	57	60
11	13	15	18	21	35	50	55	59	63	66
15	17	20	23	26	41	57	61	65	69	71
19	21	25	28	32	47	62	67	70	74	76
24	26	30	33	38	53	68	72	75	79	81
29	31	35	39	43	59	74	77	80	83	85
34	37	41	45	50	65	79	82	85	87	89
40	43	48	52	56	71	84	87	89	91	93
47	50	54	58	63	78	89	91	93	94	95
54	57	62	66	70	84	93	95	96	97	98
62	65	70	74	78	90	97	98	98.4	99	99.3
72	75	70	83	87	96	99	99.7	99.8	99.9	99.9
					Sample Size = 17					
0.03	0.06	0.15	0.3	0.6	4	13	16	20	24	27
0.6	0.9	1.5	2.1	3.2	10	21	25	29	33	36
2.1	2.7	3.8	5	7	15	28	33	36	41	44
4.3	5.2	7	8	11	21	35	40	43	48	51
7	8	10	12	15	27	42	46	50	54	57
10	12	14	17	20	33	48	52	56	60	63
14	16	18	21	25	38	54	58	62	66	68
18	20	23	26	30	44	59	64	67	71	73
22	24	28	31	35	50	65	70	72	76	78
27	29	33	36	41	56	70	74	77	80	82
32	34	38	42	46	66	75	79	82	84	86
37	40	44	48	52	67	80	83	86	88	90
43	46	50	54	58	73	85	88	90	92	93
49	52	57	60	65	79	89	92	93	95	96
56	59	64	67	72	85	93	95	96	97	98
64	67	71	75	79	90	97	98	98.5	99.1	99.4
73	76	80	84	87	96	99	99.7	99.8	99.9	99.9

Table A5. Weibull Ranking Percentiles (*continued*)
(For sample sizes of 20 or less)[a]

Confidence Level =										
0.5	1.0	2.5	5.0	10.0	50.0	90.0	95.0	97.5	99.0	99.5
Sample Size = 18										
0.03	0.06	0.14	0.3	0.6	4	12	15	19	23	25
0.6	0.8	1.4	2	3	9	20	24	27	32	35
2	2.5	3.6	4.7	6	15	27	31	35	39	42
4	5	6.4	8	10	20	33	38	41	46	49
6.5	8	10	12	14	25	40	44	48	52	55
10	11	13	16	19	31	46	50	53	58	61
13	15	17	20	23	36	51	55	59	63	66
16	18	22	24	28	42	57	61	64	68	71
20	23	26	29	33	47	62	66	69	73	75
25	27	31	34	38	53	67	71	74	77	80
29	32	36	39	42	58	72	76	78	82	84
34	37	41	45	49	64	77	80	83	85	87
39	42	47	50	54	69	81	84	87	89	90
45	48	52	56	60	75	86	88	90	92	93
51	54	59	62	67	80	90	92	94	95	96
58	61	65	69	73	85	94	95	96	97	98
65	68	73	76	80	91	97	98	98.6	99.2	99.4
75	77	81	85	88	96	99.4	99.7	99.9	99.9	99.9
Sample Size = 19										
0.03	0.05	0.1	0.3	0.6	3.6	11	15	18	22	24
0.6	0.8	1.3	1.9	2.8	9	19	23	26	30	33
1.9	2.4	3.4	4.4	6	14	26	30	33	37	40
3.8	4.6	6	8	10	19	32	36	40	44	47
6	7	9	11	13	24	38	42	46	50	53
9	10	13	15	18	29	43	48	51	55	58
12	14	16	19	22	34	49	53	57	60	63
15	17	20	23	26	40	54	58	62	66	68
19	21	24	27	31	45	59	63	67	70	73
23	25	29	32	36	50	64	68	71	75	77
27	30	34	37	41	55	69	73	76	79	81
32	34	38	42	46	60	74	77	80	83	85
37	39	43	47	51	66	78	81	84	86	88
42	45	49	52	57	71	82	85	87	89	91
47	50	54	58	62	75	87	89	91	93	94
53	56	60	64	68	81	90	92	94	95	96
60	63	67	70	74	86	94	96	97	97.6	98.1
67	70	74	77	81	91	97	98	99	99.2	99.4
76	78	82	85	87	96	99.4	99.7	99.8	99.9	99.9

Table A5. Weibull Ranking Percentiles (*concluded*)
(For sample sizes of 20 or less)[a]

					Confidence Level =					
0.5	1.0	2.5	5.0	10.0	50.0	90.0	95.0	97.5	99.0	99.5

					Sample Size = 20					
0.03	0.05	0.1	0.3	0.5	3.4	11	14	17	21	23
0.5	0.8	1.2	1.8	2.7	8	18	22	25	29	32
1.8	2.3	3.2	4.2	5.6	13	24	28	32	36	39
3.6	4.4	5.7	7	9	18	30	34	38	42	45
6	7	9	10	13	23	36	40	44	48	51
8	10	12	14	17	28	41	46	49	53	56
11	13	15	18	21	33	47	51	54	58	61
15	16	19	22	25	38	52	56	59	62	66
18	20	23	26	29	43	57	61	64	68	70
22	24	27	30	34	48	62	65	68	72	74
26	28	32	35	38	52	66	70	73	76	78
30	32	36	39	43	57	71	74	77	80	82
34	37	41	44	48	62	75	78	81	84	85
39	42	46	49	53	67	79	82	85	87	89
44	47	51	54	59	72	83	86	88	90	92
49	52	56	60	64	77	87	90	91	93	94
55	58	62	66	70	82	91	93	94	95	96
61	64	68	72	76	87	94	96	97	98	98
68	71	75	78	82	92	97	98	99	99.2	99.5
77	79	83	86	89	97	99.5	99.7	99.8	99.9	99.9

[a]For sample sizes greater than 20:

$$MR = [j - 0.30685 - 0.3863(j - 1)/(n - 1)]/n$$

where

MR = median (50%) rank ratio

j = failure rank (j = 1 for the first failure, j = 2 for the second failure, etc.)

n = sample size

BIBLIOGRAPHY

Books and Periodicals

Achieving Results with Statistical Methods, ASQC, Milwaukee, WI, 1977 (Home Study Course).

Ankenbrandt, *Maintainability Design*, Engineering Publishers, Division of A. C. Book Company, Inc., Elizabeth, NJ, 1963.

Awad, *Business Data Processing*, 5th ed., Prentice Hall, Englewood Cliffs, NJ, 1980.

Barlow, Proschan, and Hunter, *Mathematical Theory of Reliability*, Wiley, New York, 1965.

Besterfield, *Quality Control*, Prentice Hall, Englewood Cliffs, NJ, 1977.

Billington, Ringlee, and Wood, *Power System Reliability Calculations*, MIT Press, Cambridge, MA, 1973.

Bompas-Smith, edited by Brook, *Mechanical Survival: The Use of Reliability Data*, McGraw-Hill, New York, 1973.

Burns (ed.), *Advances in Reliability and Stress Analysis*, The American Society of Mechanical Engineers, United Engineering Center, New York, NY, 1979.

Calabro, *Reliability Principles and Practices*, McGraw-Hill, New York, 1962.

Coppola and Sukert, *Reliability and Maintainability Management Manual*, RADC-TR-79-200, 1979, NTIS Number AD-A073299.

Cox, *Renewal Theory*, Wiley, New York, 1962.

Davis, *Introduction to Electronic Computer*, 3rd ed., McGraw-Hill, New York, 1977.

Dhillon and Singh, *Engineering Reliability — New Techniques and Ap-*

plications, Wiley-Inter-Science Publication, A Division of John Wiley & Sons, New York, 1981.

Enrick, *Quality, Reliability, and Process Improvement*, 8th ed., Industrial Press, New York, 1985.

Goldman and Slattery, *Maintainability: A Major Element of System Effectiveness*, 2nd ed., Wiley, New York, 1967.

Green and Bourne, *Reliability Technology*, Wiley, New York, 1972.

Hahn and Shapiro, *Statistical Models in Engineering*, Wiley, New York, 1967.

Halpern, *The Assurance Sciences*, Prentice Hall, Englewood Cliffs, NJ, 1978.

Ireson (ed.), *Reliability Handbook*, McGraw-Hill, New York, 1966.

Jardine, *Maintenance, Replacement and Reliability*, Halsted Press, Wiley, New York, 1973.

Juran, *Quality Control Handbook*, 3rd ed., McGraw-Hill, New York, 1974.

Kapur and Lamberson, *Reliability in Engineering Design*, Wiley, New York, 1977.

Kimball, *Failure Analysis*, Proceedings Eighth National Symposium of Reliability and Quality Control.

Lloyd and Lopow, *Reliability: Management, Methods and Mathematics*, 2nd ed., by Author, Redondo Beach, CA, 1977.

Locks, *Reliability Maintainability and Availability Assessment*, Hayden, Rochelle Park, NJ, 1973.

Mann, Schafer, and Singpurwalla, *Methods for Statistical Analysis of Realiability and Life Data*, Wiley, New York, 1974.

McCormick, *Human Factors Engineering*, 4th ed., McGraw-Hill, New York, 1975.

Myers, Wong, and Gordy, *Reliability Engineering for Electronic Systems*, Wiley, New York, 1964.

Nelson, *Applied Life Data Analysis*, Wiley, New York, 1982.

Nonelectronic Parts Reliability Data, Reliability Analysis Center, RADC/RBRAC, Griffiss AFB, NY, 1978.

Pieruschka, *Principles of Reliability*, Prentice-Hall, Englewood Cliffs, NJ, 1963.

Rau, *Optimization and Probability in Systems Engineering*, Van Nostrand Reinhold, New York, 1970.

Roberts, *Mathmatical Methods in Reliability Engineering*, McGraw-Hill, New York, 1964.

Sandler, *System Reliability Engineering*, Prentice-Hall, Englewood Cliffs, NJ, 1963.

Shooman, *Probabilistic Reliability: An Engineering Approach*, McGraw-Hill, New York, 1968.

Smith, *Introduction to Reliability in Design*, McGraw-Hill, New York, 1976.

Military Standards

MIL-HDBK-217A, Reliability Stress and Failure Rate Data for Electronic Equipment, 1965.

MIL-HDBK-217B, Reliability Stress and Failure Rate Date for Electronic Equipment.

MIL-HDBK-472, Maintainability Prediction.

MIL-STD-439B, The Electronic Circuits, 1961.

MIL-STD-470, Maintainability Program Requirements (for Systems and Equipments).

MIL-STD-471, Maintainability Demonstration.

MIL-STD-721B, Definitions of Effectiveness Terms for Reliability, Human Factors and Safety.

MIL-STD-781B, Reliability Tests: Exponential Distribution.

MIL-STD-785A, Reliability Program for Systems and Equipment Development and Production.

MIL-STD-790C, Reliability Assurance Program for Electronic Parts Specifications.

MIL-STD-1472, Human Engineering Design Criteria for Military Systems, Equipment and Facilities, 1968.

INDEX

303